MW00639920

THE
SECRET
LIFE
OF THE
UNIVERSE

An ASTROBIOLOGIST'S SEARCH
for
the ORIGINS *and* FRONTIERS *of* LIFE

NATHALIE A. CABROL

SCRIBNER

NEW YORK   LONDON   TORONTO   SYDNEY   NEW DELHI

Scribner
An Imprint of Simon & Schuster, LLC
1230 Avenue of the Americas
New York, NY 10020

Copyright © 2023 by Éditions du Seuil

Originally published in France in 2023 by Éditions du Seuil, S.A
as *À l'aube de nouveaux horizons*

First Scribner hardcover edition August 2024

SCRIBNER and design are trademarks of Simon & Schuster, LLC

Simon & Schuster: Celebrating 100 Years of Publishing in 2024

For information about special discounts for bulk purchases,
please contact Simon & Schuster Special Sales at 1-866-506-1949
or business@simonandschuster.com.

The Simon & Schuster Speakers Bureau can bring authors
to your live event. For more information or to book an event,
contact the Simon & Schuster Speakers Bureau at 1-866-248-3049
or visit our website at www.simonspeakers.com.

Interior design by Kathryn A. Kenney-Peterson

Manufactured in the United States of America

10 9 8 7 6 5 4 3 2 1

Library of Congress Control Number: 2023059688

ISBN 978-1-6680-4668-5
ISBN 978-1-6680-4670-8 (ebook)

In Memoriam
Edmond A. Grin
(January 26, 1920–August 17, 2022)

Who knew that thirty-six years would go by so fast. We did everything together, and most would say today that they could not think of one without thinking of the other; and you could bet that when you saw one, the other was never very far away. We were a team, soulmates, married, and best friends, maybe simply one spirit in two bodies. Together, we climbed mountains, explored deserts, even Mars. We could also simply stand in each other's arms, silent, gazing at the ocean or at the starry night on top of heavens-high volcanoes in the Andes. You transformed the shy young woman I was into a leader and a volcano diver, and you smiled when I donned my suit to dive in the Licancabur Lake at almost six thousand meters elevation. I do not know where my life's journey will take me from this point on, but what I know is that, wherever I go, it will bring me back to you.

CONTENTS

CONTENTS

CONTENTS

CONTENTS

THE
SECRET
LIFE
OF THE
UNIVERSE

INTRODUCTION

On July 11, 2022, the James Webb Space Telescope (JWST) returned its first images, penetrating the wall of time to show us the universe just a few hundred million years after its formation. In a marvelous cosmic irony, this immersion into the depths of our origins propels us into the future, where a revolution looms large in astronomy, in cosmology, and in astrobiology—the search for life in the universe. JWST comes after a few decades of space and planetary exploration during which we have discovered countless habitable environments in our solar system—for (simple) life as we know it, but also thousands of exoplanets in our galaxy, some of them located in the habitable zone of their parent stars. We are living in a golden age in astrobiology, the beginning of a fantastic odyssey in which much remains to be written, but where our first steps bring the promise of prodigious discoveries. And these first steps have already transformed our species in one generation in a way that we cannot foresee just yet.

I was born barely two years into the Apollo program and yet, unlike my parents, watching the Earth from space was already in my DNA. It is as natural for me to see our planet rise in an alien sky as it was for them to see the moon rise over the horizon. I embraced my life's calling and made astrobiology my forever passion, keeping my eyes and mind squarely set on the stars as I explore a backyard of already familiar planetary landscapes. In my mid-twenties, just a few days shy of my birthday, Voyager 2 zoomed by Neptune. The spacecraft flew barely 4,800 kilometers above the planet's north pole to return historical images that made the front page of newspapers worldwide. It was the farthest humanity had ever reached, albeit through the means of a robotic probe. That would be the last planetary encounter for Voyager 2 before the spacecraft finally left the heliosphere forever, only six years after its twin, Voyager 1. When we think of it, only sixty-seven years went by since Sputnik started to circle Earth's orbit and the moment the last of the two Voyagers sailed outside the confines of our solar system—a blink of a cosmic eye. Humanity, with increasing curiosity, was just starting to open its eyes to its planetary neighborhood.

In 1996, I saw the first astrobiology workshop at NASA Ames Research Center in California, where I was a young researcher. Two years later, I witnessed the birth of the NASA Astrobiology Institute (NAI). Missions were showing us that the solar system was populated by worlds where simple life could have been possible in the distant past and maybe could still be today. Astrobiology was born from these explorations and today it is at the heart of all spacefaring nations.

Shortly after the NAI was founded in 1998, and just a few kilometers away, I entered a small conference room at the SETI (Search for Extraterrestrial Intelligence) Institute in Mountain View one late-summer afternoon. I knew that the institute's mission was to lead humanity's quest to understand the origin and prevalence of life and intelligence in the universe and share that knowledge with the world. It represented everything I was passionate about. My heart was pounding in my chest. On the opposite side of the table was astronomer Frank Drake, a gentle giant in the field, who invited me to sit down. In a soft voice, he kindly inquired about my experience and research. Then he asked me how they related to his famous "Drake equation" and the programs for the search for life in the universe at the institute, which span between the origins of life and the search for extraterrestrial intelligence. Christopher Chyba, then the director of the Carl Sagan Center, the research wing of the SETI Institute, was out of town that week. So it was Frank himself who interviewed me and who accepted my application to become a planetary scientist at the institute.

In 2003, as the coinvestigator of a team supported by the NAI, I began my personal scientific odyssey to understand planetary habitability and the type of life that could have once existed on early Mars and possibly still today. This became the High Lakes Project, and with my team, we started to explore unique analogs to the red planet in extreme terrestrial environments in the Andes. We also began to document how life adapts, what its geological and biological signatures are, how rapid climate change impacts lake ecosystems and habitats, and the relevance of all of these issues to planetary exploration.

Nearly two decades after I became a part of the SETI Institute, it was my turn to lead a new NAI-funded project. With a multidisciplinary team, our goal was to develop techniques for planetary exploration that could enable the detection of the signatures of life, called biosignatures. And 2015 was also the year when I was appointed as the director of the Carl Sagan Center, becoming the SETI Institute's chief scientist.

Thinking about the history of the SETI Institute, the key role played by Carl Sagan in founding astrobiology and his support of the SETI search, I realize this is, without any doubt, the greatest honor I have ever received. But one of my mottos in life is that titles are only as good as what we do with them. In that spirit, one of my first orders of business was to hang Carl Sagan's portrait next to my desk to make sure it

3

would be the first thing I'd see when I arrived in the morning. Just as a reminder. I chose a black-and-white photo, showing Carl with his chin resting on his palm, maybe a bit reflective, but with a kind and focused gaze. Although he left us years ago, he remains an endless source of inspiration to all of us astrobiologists. I also attached a deeper meaning to it. First, it was a nod to my past and a bridge to my early years in France, when Carl stirred my passion for astrobiology with his words and his fire during our meeting in Paris in November 1986. It is also a constant reminder of Carl's urging that we should never be afraid to think bold thoughts: "We can judge our progress by the courage of our questions and the depth of our answers, and our willingness to embrace what is true rather than what feels good." Of course, this quote has to be immediately followed by his belief that "extraordinary claims require extraordinary evidence." Known as ECREE, or the "Sagan standard," it ensures that our ideas pass the smell test. Carl's standard, which echoes the thinking of some of the greatest minds, like Flournoy, Laplace, Hume, and Jefferson, is a necessity in science at each step, and it is an absolute must in our quest for life in the universe.

In December 2016, I was standing at the podium of a large conference room in the Moscone Center in San Francisco, where hundreds of my colleagues had gathered. The American Geophysical Union honored me that year with the presentation of the Carl Sagan Lecture. I had butterflies in my stomach. As I stood at the podium, I decided to start by mentioning our 1986 encounter. I spoke about how momentous this had been for me. Carl seemed to so uncannily show up at the most unexpected times and in the most unexpected ways in my life, even long after his death. In our brief time together in Paris, he told me about passion, resilience, and never letting anybody define what I should do or who I should become. I smiled when he insisted that I always be true to science and data, not opinions, even when they go against the establishment.

That day, he did not give me any advice about going to this or that fancy school or asking for support from a big lab. Instead, his were words for life that would serve me anywhere I would go. He went on about his own passion for a brief moment, and there he was in front of me, the great Sagan; even better, just for me.

Preparing this lecture made me reflect on how much has changed since Carl's vision of "the shores of the cosmic ocean" in the eighties. This idea was introduced in an episode of the TV series *Cosmos*, where he examined Earth's place in the universe by taking us on a journey of exploration into space. And here, the answer is ambiguous. What has dramatically changed is the number of missions, the amount of data, and the technology, systems, and instruments we can apply today to the search for life in the universe. In barely a few decades, we saw dizzying progress supported by constant innovation exponentially advance our knowledge and provide more and smarter ways to explore. And we now have astrobiology, a scientific field dedicated to this search. The vast volumes of data returned from planetary and space exploration since the seventies called for a multidisciplinary approach and a holistic perspective to their interpretation. Astrobiology provided this new platform for the search for life in the universe and has since revolutionized how we do science, by bringing together distinct domains of inquiry, their perspectives and methodologies, and demonstrating elegantly that discoveries and innovation lie at the nexus of disciplines.

But while exploration brings an abundance of data, we seem to be generating new intellectual frameworks at a slower pace. Our vision of the universe and life's potential has been completely transformed in less than forty years; however, fundamental questions regarding life's origins and nature remain unanswered. Maybe part of the challenge resides in the fact that we are both the observer and the observation; we are life trying to understand itself and its origin.

Today, we might still not know exactly where we are heading and what we are looking for, but it does not really matter. Answers will present themselves as we go. What truly matters is that we have set sail. We are now on our way on the most remarkable journey humanity has ever undertaken, searching for our origins and for a cosmic echo that will finally tell us one day that we are not alone.

This is the odyssey that I wish to share with you in the following pages—the discoveries, of course, but also the questions that remain unanswered, including that of our humble beginnings: What is the origin of life? What is life? Are we alone in the universe?

This is an epic quest that seeks answers to the mysteries of life, a balancing act of microscopes exploring the most elemental bricks of life and radio telescopes searching for advanced extraterrestrial civilizations in the farthest reaches of the universe. This voyage of incredible significance sparks awe, curiosity, and an insatiable thirst for understanding. And it brings necessary changes in perspectives. We start thus by reflecting on the meaning of the Overview Effect experienced by astronauts who ventured to the moon, and into Earth's orbit, and returned profoundly transformed. From their unique vantage point, they gazed upon our "Pale Blue Dot" suspended in the vastness of space. There, they saw the fragility of our planet and the dire need for humanity to start thinking as one, the destiny of all people of Earth irremediably sealed in one ephemeral sea of blue. This apparent fragility is also our strength and a beacon lighting up our way toward a brighter future of togetherness. And it is the source of inspiration for our most profound questioning, making it even more important to understand who we are, where we are coming from, and if others out there are wondering about the same questions while looking up in their own night sky.

We next embark on a quest to uncover the origins of life, pondering the mysteries of Earth's early history. Today, new biogeochemical and biophysical theories challenge past models, pushing the boundaries of our understanding and redefining the conditions for life's inception. With these new perspectives, we set out to understand if the conditions for life, and maybe life itself, would have been possible beyond Earth in the past of our solar system. Could it still be possible today? Finding even the smallest of microbes on another world would be incredibly important for what it would foretell for the potential for life in the universe. And the more we advance, the more the possibilities broaden. The hellish world of Venus, still shrouded in secrecy, beckons us to unveil its mysteries, which it may soon reveal. The debate over the presence of phosphine gas in its atmosphere and hints of active volcanism pose questions that cannot be ignored and that upcoming missions will soon address. Mars holds an enduring fascination. Recent missions have demonstrated that it was once a habitable planet and suggest that it could still harbor habitable environments at depth. From its early exchange of materials with the Earth, it also raises a tantalizing question: Could life on Earth have Martian origins? Farther yet into the solar system, ocean worlds capture our imagination with their potential for life lurking beneath the icy moon surfaces. From the frozen expanses of Enceladus to Europa and their subsurface oceans, we explore alien aquatic realms in search of signs of life. Of them all, Titan is the most exotic world and our best shot at studying processes that could lead to life as we do not know it in our own solar system. The Dragonfly mission will aim to reveal the secrets of this enigmatic world.

Then we go beyond our own neighborhood to discover other planetary systems and explore exoplanets. The revelations of the Kepler mission were just the beginning. The seminal mission has since been followed by many more advanced observatories on the ground and in space, opening new frontiers in our search for Earth-like worlds in our galaxy. A multitude of exoplanets entice our curiosity, often blurring the lines between science and science fiction. We seek Earth 2.0 among the stars, revisiting Carl Sagan's vision of an expansive universe. And, in the pursuit of technologically advanced extraterrestrial life, we explore the Drake equation. The Fermi paradox challenges our expectations, leading to radical hypotheses, understanding that, ultimately, they might be simply a reflection of our own limited capacity to think outside the human and terrestrial box. But, relentless, our quest extends across galaxies as we seek signs of advanced extraterrestrial civilizations with new SETI projects scanning the cosmic ocean for signals.

In our concluding chapters, we confront the definitions of life and artificial intelligence. We traverse ethical dilemmas and contemplate the possibility of synthetic life. We finally reflect on the pathways and destinies of life's journey through the cosmos.

As we set our course for interstellar exploration, we are reminded that the universe is both an enigmatic puzzle and a profound mirror reflecting our own existence. And this is how the search for life beyond Earth, a quest that began with a simple perspective-altering moment, has now evolved into a profound exploration of our place in the universe.

The Earth viewed from the Apollo 8 spacecraft as astronauts circle the moon for the first time on December 14, 1968.

1

THE OVERVIEW
EFFECT

A stronauts often describe an overwhelming feeling of awe as they watch the Earth from above. They call it the Overview Effect, an expression that tries to capture the profound emotional impact of seeing our planet suspended in the darkness of space with only a thin atmosphere for a protective shield. From this vantage point, political borders disappear. The tumult of human hustle vanishes in a deafening silence to give way to the realization that we are all aboard the same vessel in space, united by the same fate. Any other consideration is simply irrelevant. Most of these space wanderers come back to Earth with a heightened sense of responsibility toward our environment and an acute understanding of the vulnerability of life, which so profoundly contrasts with how our civilization treats them both down here. Many feel compelled to share their experience; some become activists. The Overview Effect is possibly the most profound legacy our generation will leave to those who will follow. It marks a turning point in humanity's history, a change in perspective, one that comes at a pivotal time when it is up to us to decide whether we want to drive ourselves into oblivion or soar into the vastness of space.

Thinking of our place in space, of whether we are alone in the universe or what is the origin of life, is as old as humanity. However, our generation is the first to see Earth from above and beyond its atmosphere. With that perspective comes new knowledge—and a few messages for the ages. The first one was delivered by astronauts making their way to the moon, speaking of our planetary cradle and humanity's place of origin. The oneness of a fragile biosphere and our responsibility as the dominant species were then revealed for the first time.

A PALE BLUE DOT

Other overview moments took place in the decades that followed, and their messages were delivered from much farther away. On February 14, 1990—ten months before the Galileo experiment, Carl Sagan asked to turn the Voyager 1 camera back toward the Earth. The spacecraft was by then already 6 billion kilometers away. In an image forever immortalized as "the Pale Blue Dot," the Earth appeared no larger than a pixel. Carl described it as "a mote of dust suspended in a sunbeam." The Pale Blue Dot showed our fragility and the need to preserve "the only home we've ever known." It illustrated our togetherness, "our responsibility to deal more kindly with one another," and the necessity for us to stop fighting for a mere "fraction of a dot." The Voyager mission also gave us a family portrait, the first perspective of our place within our planetary family in the solar system's neighborhood.

Other spacecrafts have looked back at the Earth and the moon since, an unlikely duo in a sea of darkness. But the vision of the Pale Blue Dot will forever remain the first one. It was also the prelude to something much bigger. An astronomical revolution was about to unfold barely a few years later and extend far beyond our neck of the galactic woods. In January 1992, astronomers Aleksander Wolszczan and Dale Frail announced the discovery of two planets orbiting a pulsar (PSR B1257+12) some 2,300 light-years away. These were the first planets beyond our solar system, or "exoplanets," to be found. In 1995, Michel Mayor and Didier Queloz discovered 51 Pegasi b, an exoplanet orbiting a sun-like star fifty light-years away. Over twenty-five years later, space telescope missions like Kepler, TESS (the Transiting Exoplanet Survey Satellite), Hubble, and now the James Webb Space Telescope (JWST) have helped us discover and confirm over five thousand exoplanets. Some of them are Earth-like, and a fraction of them sit in the habitable zone of their parent stars. But it is not all about space technology. Cutting-edge ground-based telescopes significantly contribute to these discoveries through direct imaging and spectral analyses as well. In their field of view, exoplanets, just as our Pale Blue Dot was with Voyager, are no bigger than a pixel.

The Overview Effect, wherever it happens, is about a change in perspective, one from which there is no turning back. Copernicus taught us long ago that the Earth was neither at the center of the universe nor the center of the solar system, for that matter. We also learned from the work of Harlow Shapley and Henrietta Swan Leavitt that the solar system does not even occupy any particularly prominent place in our galaxy. It is simply tucked away at the inner edge of Orion's spur in the Milky Way, 27,000 light-years from its center, in a galactic suburb of sorts. Our sun is an average-sized star located in a galaxy propelled at 2.1 million kilometers per hour in a visible universe that counts maybe 125 billion such cosmic islands, give or take a few billion. In this immensity, the Kepler mission taught us that planetary systems are the rule, not the exception.

ARE WE ALONE IN THE UNIVERSE?

This is how, in a mere quarter of a century, we found ourselves exploring a universe populated by as many planets as stars. Yet, looking up and far into what seems to be an infinite ocean of possibilities, the only echoes we have received so far from our explorations have been barren planetary landscapes and thundering silence. Could it be that we are the only guests at the universal table? Maybe. As a scientist, I cannot wholly discount this hypothesis, but it seems very unlikely and "an awful waste of space," and for more than one reason.

To begin with, the elementary compounds that make the life we know, carbon, hydrogen, nitrogen, oxygen, phosphorus, and sulfur (CHNOPS for short), are common in the universe. It is no accident that we are made of them. They are the star stuff Carl always talked about. Organic molecules and volatiles are found at the surface of Mars, in the geysers of Saturn's tiny moon Enceladus, in the atmosphere of Titan, in Triton's stratosphere, and on comets. We also discovered them on asteroids, not to mention dwarf planets Ceres and Pluto, and these are only a few examples. Much farther away still, nearly two hundred types of prebiotic complex organic molecules were detected in interstellar clouds near the center of our galaxy. They included the kinds that could play a role in forming amino acids—the building blocks of the life we know. Granted that organic molecules are not life, but they are the elemental building blocks life uses for its carbon and hydrogen backbone, and they are everywhere.

The sheer number of possibilities adds to the probability of an abundance of life in the universe. A basic extrapolation of the Kepler data on the number of exoplanets in our galaxy alone suggests that tens of billions of Earth-sized planets could be located in the habitable zone of sun-like stars. Suppose only one in a billion has developed a type of life that made it to higher levels of complexity and intelligence. Then nearly a dozen advanced civilizations could populate our galaxy alone. Even if it were only one per one hundred galaxies, there could still be billions of them in the universe. And because the probability distribution in nature predicts more puddles than large lakes, more small buttes than Himalayas, more small planets than large ones, and more simple life than complex life, it follows that the universe is likely teeming with planets harboring simple life. What precedes is an obvious oversimplification, but it is not an unreasonable one, and there are several possible scenarios.

The Earth might not be a gold standard for how rapidly life develops. One the one hand, it could represent a population of relatively slow planetary bloomers. After all, it took over 80 percent of our planet's geological evolution to reach complex life. On the other hand, it could be an example of life on a universal fast track. Living organisms possibly left indirect traces of their presence on our planet in the few oldest rocks that still exist, which formed less than a couple of hundred million years after the Earth's crust had cooled. In truth, we do not know any better because we only have one data point, and that's us. Everything is relative and depends on what type of life we refer to. Our knowledge is still modest and imperfect, particularly for these deep times of early Earth, since plate tectonics and erosion have destroyed most of the geological record. Further, life could also be the result of a generational process associated with the formation of specific stars—in our case, sun-like stars.

Our galaxy is about 13.6 billion years old, formed barely 200 million years after the Big Bang, but it did not produce sun-like stars right away. The oldest (Population III) were short-lived (2 to 5 million years), massive, luminous hot stars that would have existed very early in the universe. They had virtually no metals (elements other than hydrogen or helium) in their composition. Their existence remained hypothetical for a long time, only inferred from indirect observations of a galaxy seen through gravitational lensing in a faraway region of our universe. Recent observations with Gemini North, a ground telescope, and the James Webb Space Telescope seem to confirm their past existence and hint at titanic stars, hundreds of times more massive than our sun.

Population II stars are more recent and metal-poor in comparison to younger stars like ours, for instance. They are distributed between the bulge near the center of our galaxy and its halo. The death of these Population II and III stars produced the heavier elements now being used by life as we know it. Population I, or metal-rich stars, are the youngest, and our sun is part of that population. The biogenic elements that make life on Earth are the most abundant in the universe and on our planet. The exception is phosphorus, which could have been delivered to Earth's early atmosphere by extraterrestrial material. It could have been incorporated into Earth during accretion and the Late Heavy Bombardment period through impacts with asteroids and comets. Phosphorus was repackaged into useful forms for biology through chemical reactions and became an essential component of the structural backbone of our genetic code. It drives the energy behind nearly all of life's metabolism.

The universe has produced biogenic elements for a very long time, as demonstrated by JWST, with the discovery of complex organic molecules in a galaxy more than 12 billion light-years away! Still, life developed on Earth, and maybe elsewhere, possibly because they became sufficiently abundant with the most recent Population I stars, like our sun. If true, this could make life a process associated with specific generations of stars. On the other hand, these biogenic elements are so old that they had to experience a long and complex chemical history before being incorporated into the Earth's biochemistry, a transformative pathway that could also be key to the origin of life. We do not know if this history played a role in the origin of life on Earth. But if life as we know it is indeed associated with the birth of specific stars, then the universe could just be starting to blossom with cradles of life, the kind that might be chemically and biologically familiar to us.

COEVOLUTION: A KEY PRINCIPLE

Critically, this relatively quick and dirty back-of-the-envelope estimate only considers life as we know it and Earth-like planets as the only worlds with the potential to develop life, but nothing at this point allows us to declare that life's inception should be limited to these criteria. Decades of exploration in space and extreme terrestrial environments have broadened our search horizons rather than narrowed them down. They have helped us refine our understanding of planetary habitability and environmental habitats. They also emphasized the role of a fundamental concept in astrobiology: the coevolution of life and its environment.

Earth's biological and environmental evolution are intertwined and inseparable, and environmental changes have accompanied life since its very beginnings. Today, we take this notion for granted, but that was not always the case. Barely three centuries ago, we still believed that the environment stayed the same over time, and species were adapted once and for all. Then, in the nineteenth century, the naturalist Charles Darwin proposed that the environment was ever-changing and forced life to adapt or disappear. Organisms that adjusted to their environments were the most successful in surviving and reproducing. That's the famous "survival of the fittest." Today, our understanding is that changes between environment and life are directional and interdependent. In other words, environmental changes accompany changes in the history of life, either as causes or as effects, and they take place at many different scales in space and time.

At large temporal scales, the habitability of our planet is closely connected to the evolution of the sun's activity, which varies over time. Our sun formed 4.6 billion years ago and went through early birth pangs, blasting intense solar winds for a few million years before entering its main sequence, a phase when a star is fusing hydrogen in its core and the outward pressure from core nuclear fusion and the inner push from gravity are balanced. It is about halfway through its life now. Its luminosity might have been only 75 percent of today's during the Archean eon, 2.5 to 3.8 billion years ago. In theory, temperatures would have been too low at that time to allow liquid water at the surface of the Earth, and liquid water is central to life's development and survival. Yet, the earliest fossils show that life took hold anyway. If our planet had fewer continental masses, it would have reflected less light into space than it does today. More oceans would have absorbed more sunlight and stored more heat, thus possibly compensating for a fainter sun. The Earth's atmosphere in the Archean was also rich in carbon dioxide and methane, two powerful greenhouse gases that could have helped maintain the surface temperature between 0 and 40 degrees Celsius.

By contrast, toward the end of the sun's life, nearly 7 billion years from now, it will enter a red giant phase. Its gas envelope will expand past the orbit of the Earth, possibly Mars. But life on Earth will run into trouble long before its star incinerates it. As the sun evolves, its brightness increases by 10 percent every billion years. This process will lead to changes in the solar system's habitable zone, a region defined as a range of distances away from a star where liquid water can be stable on the surface of a planet. An increase of 10 percent means temperatures will become too hot for liquid water within the next billion years. The oceans will then start to evaporate. As they do, more water vapor will accumulate in the atmosphere, acting as a greenhouse that will accelerate the rise in surface temperatures and trigger even more evaporation. Our planet will then face a runaway greenhouse effect from which life ultimately will not recover. Closer to the sun, Venus has already faced this scenario.

One billion years is both far into the future and the blink of an eye at a geological scale. Life's odyssey on Earth is already 4 billion years old, or about 80 percent into its possible residence time on this planet. Today, the sunlight is just right to support life, but it is not the only source of energy. The Earth itself provides a range of free energy, including geothermal and geochemical energy, oxygen, and radiolysis, the latter producing elemental hydrogen and oxidants that some microbes use as nutrients. These energy sources mediate the diversity and complexity of living organisms. In turn, life develops, and its increasing diversity and complexity affects the Earth's carbon cycle, climate, ecosystems, and the oxygen level in its atmosphere. Changes in biodiversity are thus critical in shaping our planet's geochemical cycles.

Mutations are another driver of change for life, and, in turn, its environment. They modify the genetic sequence of organisms and are an indispensable requirement for life's evolution. In most cases, they appear spontaneously without a specific cause being identified. As they shape life's diversity, gene mutations also impact biodiversity and planetary habitability.

Other examples of how life modifies its environment can be found at both ends of the temporal scale of evolution with planetary-wide repercussions. When the Earth formed, its primordial atmosphere lacked free oxygen. A couple of billion years later, blue-green algae started to photosynthesize using sunlight, water, and carbon dioxide to produce carbohydrates and oxygen. Once the biological production of oxygen began, it took an additional few hundred million years for its accumulation in the atmosphere. The air we breathe today results from life changing the environment.

Meanwhile, we are currently the most visible illustration of this process at the other end of Earth's history. As a result of human activity, the global average concentration of carbon dioxide has increased by 12 percent since 2000. Likewise, the concentration of methane has more than doubled since preindustrial times. Increases in greenhouse gases are associated with global warming and changes in habitability worldwide, loss of countless species across the entire biosphere, and a severe alteration of the ecosystem. In a twisted irony, the most advanced species on this planet is methodically cutting the branch it sits on, and knowingly destroying the environment responsible for its rise and its development.

NATURAL CYCLES AND COSMIC THREATS

Earth's natural cycles, environment, and physical and chemical characteristics were initially dictated by the composition of dust, gas, and ice from which it was born, the distance of our planet to the sun, and other astronomical parameters. Here, three aspects of Earth's orbit around the sun play a fundamental role in the evolution of our planet's habitability. The first is the shape of Earth's orbit, which deviates from a perfect circle; the second is the angle of the Earth's axial tilt. The third is the wobble of the Earth as it rotates and is being pulled by tidal forces generated by the sun and the moon's gravity. These cycles affect the amount of energy that the surface of the Earth receives from the sun. They are known as the Milankovitch cycles[1] for the Serbian astrophysicist who developed the theory.

Individually and together, they influence long-term climate patterns over tens to hundreds of thousands of years, driving ice ages and interglacial periods. These natural cycles are sometimes perturbed by external forces. Random events reset the clock for both environment and life over periods ranging from years to millions of years. They may be cosmic in origin, like asteroids and comets that collided with our planet regularly throughout its geological history and have redirected life's evolution more than a few times. If we look at Earth's impact cratering record over the past 260 million years, a peak appears every 27.5 million years that aligns well with a known cycle of mass extinctions. Five of the six largest impact craters identified from this extended period, including the one attributed to the extinction of the non-avian dinosaurs, were formed within the time frame associated with mass extinctions. Some see a cause-and-effect relationship in these observations. Every 27.5 million years, our solar system passes through our galaxy's dense mid-plane. At

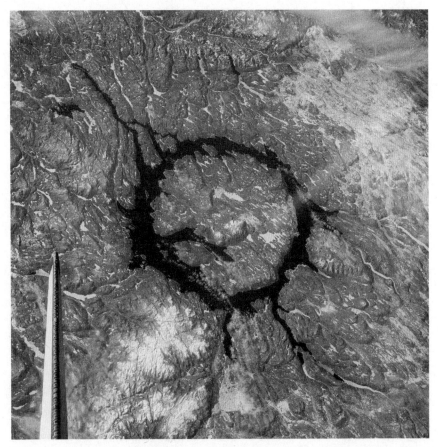

The Manicouagan crater in Quebec, Canada, is a large 100-kilometer-wide impact crater formed around 212 million years ago by an asteroid. It has been shaped since by glacial activity and is now a reservoir drained by the Manicouagan River, flowing into the St. Lawrence River.

these times, gravitational tugs-of-war may expel comets from the Oort cloud, where they otherwise reside, and send them on a collision course with the inner planets. Wandering stars passing close to the Oort cloud would have the same effect, but with much less predictability.

So far, in recent times, we have been surprised only by asteroids of relatively small sizes, but the Chelyabinsk airburst on February 15, 2013, with its explosion and debris from a seventeen-to-twenty-meter asteroid, was enough to wreak havoc all over town. On June 30, 1908, the Tunguska event was also attributed to an airburst from a 200-meter stony meteoroid hurtling through the Earth's atmosphere at 72,500 kilometers per hour. In the process, it leveled 80 million trees over a 2,150 kilometer area of the forest without even leaving a crater, since it exploded high in the atmosphere.

Future impacts of asteroids and comets are not a matter of if, just a matter of when, and, in this department, there is good news and there is bad news. The good news is that the bigger asteroids and comets are, the fewer their number and the less frequent the risk of impact. The bad news is that it only takes one to trigger a mass extinction event that we are neither immune to nor prepared for. We already know of potentially dangerous near-Earth objects, like the 2.4-kilometer-wide asteroid Toutatis or the 365-meter-wide Apophis predicted to pass barely 30,600 kilometers away from the Earth on April 13, 2029, closer than many of our geostationary satellites. And for those who are superstitious, yes, that will be on a Friday.

Understanding and modeling the trajectories of these objects and characterizing their size, shape, mass, composition, and rotational dynamics is critical. It helps us forecast the risks they represent and may give us enough time to react. To that end, new initiatives are being pursued to develop a planetary defense program to find and track these near-Earth objects that pose a threat to the Earth and find solutions to either destroy them before they destroy us or modify their trajectories. NASA's Double Asteroid Redirection Test (DART) mission represents a first attempt at understanding how we can deflect asteroids by crashing a spacecraft into one to change its motion in space. The target of this mission was the 170-meter-wide Dimorphos, a moonlet orbiting near-Earth asteroid Didymos nearly 10 million kilometers away from Earth. The spacecraft reached the asteroid on September 26, 2022, and struck it with remarkable precision, and the orbital period of Dimorphos around Didymos was altered by two minutes by the impact. The European Space Agency's mission Hera will conduct a follow-up investigation of the dual asteroid system in a few years. The Chinese National Space Administration (CNSA) might also launch a kinetic asteroid defense mission on an unspecified target as part of an effort to build a Near-Earth Objects (NEOs) defense system and increase our ability to monitor and catalog them and be in a position to provide early warnings and responses.

Comets and asteroids are only one of many other potential cosmic threats to life on Earth, as shown by the Hangenberg event. Toward the end of the Devonian period, 359 million years ago, all placoderms (armored fishes with the front part of their bodies encased in broad flat bony plates) and nearly 70 percent of Earth's invertebrates died off. Fossil spores of that period show traces of substantial damage from ultraviolet (UV) radiation, the telltale sign of a long-lasting disruption of the ozone layer. While climate change and volcanic activity could contribute to damaging the ozone layer, their effect is usually short-lived, and the oceans should have provided an effective shield against UV. What struck the Earth then had to be capable of destroying part of the ozone layer and penetrating deep underground and into the oceans. The leading suspect in that case is a supernova event powerful enough to blast the Earth's atmosphere and hit its surface with high-energy particles, leaving a lingering effect for up to one hundred thousand years. There is currently no threatening star within a distance that could lead to a biological cataclysm of this magnitude. While we might see Betelgeuse go supernova in our lifetime, this red supergiant located 642 light-years away from Earth is too far to present any danger. Local gamma-ray bursts directed at the Earth could have similar extinction-level results, but so far, their sources have been located billions of light-years away and only outside of our galaxy.

Geological and environmental catastrophes have also profoundly affected life's evolution. Some were sudden and violent; others disrupted the environment over extended periods. A classic example is the Deccan Traps, which started to form 66 million years ago through a series of pulsing volcanic eruptions that lasted over thirty-three thousand years during the Cretaceous period and correlate well with significant climate variations and mass extinctions. The wandering of plate tectonics may also be responsible for the Ordovician extinction 445 million years ago. As the supercontinent Gondwana moved into the Earth's southern hemisphere, sea levels fluctuated over millions of years, destroying countless life habitats and eliminating 85 percent of all species. These are only samples of how the environment also shapes life's evolution.

Over eons, life had to adapt or perish. It battled countless assaults from a changing environment following random events and natural cycles of glacial ages, atmospheric and oceanic chemistry changes, and decline in ocean oxygen levels. The sum of knowledge we accumulate through the study of the Earth gives us an ever more profound understanding of the coevolution of life and its environment, of what habitable zone and habitability mean at a planetary scale, and how they can fluctuate over the course of a planet's history.

Meanwhile, as spacecrafts roamed the solar system in the past decades, the scientific exploration of our world intensified in parallel, reaching into its most extreme environments. That exploration brought yet again a change in perspective. Everywhere we looked, from the highest mountain to the deepest abyss, in the most acidic or basic environments, the hottest and coldest regions, in places devoid of oxygen, within rocks—sometimes under kilometers of them—within salts, in arid deserts, exposed to radiation or under pressure, life was still present. And it told us tales of possible places of origin and life's extraordinary ability to adapt to the most extreme conditions.

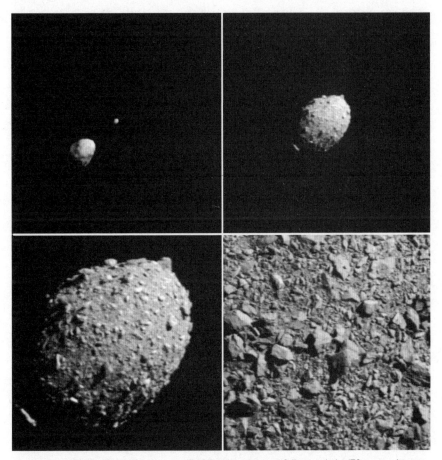

After a journey of 10 million kilometers, the DART mission successfully struck the 170-meter-diameter Dimorphos with head-on precision. The image sequence shows DART nearing impact, Dimorphos covered in impact debris and resembling Bennu and Ryugu, and the probe's final approach before hitting the surface. DART covered the last ten kilometers in just one second.

As images and data streamed back from faraway planets and moons, we realized that some of their ancient and present environments offered close analogies with these terrestrial extremes. From that moment on, the exploration of our planet would not only help us piece together the origins and the coevolution of life and its environment here on Earth. It would also add to our vocabulary the notion of environmental habitability, which shows that a planet or a moon can be located well outside our sun's habitable zone and yet, still be habitable. And that's how Earth's exploration opened up alien worlds of possibilities for the search for life in our solar system and beyond.

2

SPARKS
OF LIFE

Today, thanks to the search for exoplanets, we know of at least 3,800 stars with planets orbiting them, and that's in a very small region of our galaxy. Over a third of them have at least three planets. Our sun, Kepler-90, and TRAPPIST-1 have between seven and eight planets, and Tau Ceti may have up to ten. Our solar system is only one of countless planetary systems in our galaxy. Still, we are unique in many ways, first and foremost in how stardust assembled to make us who we are: complex and sentient actors and spectators of an infinite cosmic show. But unique does not have to mean alone, and, to finally answer whether life is a common or rare occurrence in the universe, we must first understand how it emerged here on Earth.

THE ORIGIN(S) OF LIFE ON EARTH

Life's origin is still uncertain today and probably the most fundamental and fascinating question of astrobiology for its scientific and philosophical implications. It is also central to any consideration of extraterrestrial life because how we search for life beyond our planet is guided, and possibly limited, by what we know from Earth. But this is all we have to get started on our quest.

Recent studies have proposed that biogenic carbon—carbon that has been produced by life—may be present in 4.1-billion-year-old rocks from Jack Hills in Western Australia. Tiny fossils, possibly as old as 4.28 billion years, have been found in remnants of an ancient oceanic floor in Canada. The confirmation of these two discoveries would be profound. It would tell us that simple, albeit already complex, life was present on Earth barely 200 million years after the surface of our planet had solidified. To date, the oldest evidence of life has been produced by seafloor hydrothermal vents, shallow marine continental margins, and intertidal environments. There are also microfossils and rocky formations created by the interactions of microbes with their environment. Isotopic and molecular biomarkers show that single-celled organisms were already abundant at the dawn of time on our planet. Hot spring and vent deposits constitute the earliest evidence of life on land 3.5 billion years ago.

An active hydrothermal vent at El Tatio in the Chilean Andes, a type of environment analogous to those that preserve the earliest fossils on Earth.

They are close in age to the Strelley Pool fossils that formed in a shallow sea 3.43 billion years ago in Australia. This chronology reveals that life appeared quickly on Earth. The few fossils of these early times give us a glimpse into what life was then, or in other words, the type of simple microbial organisms that existed, descendants of precursors that had made it through the hellish conditions of the late Hadean eon, a time when our planet was still cooling down and volcanic activity was widespread. The Earth then was still bombarded by enormous asteroids and comets, some reaching tens to hundreds of kilometers wide. However, we are still trying to grasp *how* it happened and what led to the dawn of life.

The *how* of the origin of life is approached through various theories. One of them is that life results from a supernatural event, which is the foundation of religions. It is generally considered outside the scientific description of nature. But, beyond their differences, science and religions are confronted with similar paradoxes and questions, such as what came before the Big Bang? What came before God? Why does the universe or God even exist, since there is no necessity for them or anything at all? These are profound questions that each of us approach in a personal way. But to put it in simple terms, science tries to understand nature by measuring it, whereas spirituality tries to understand what can't be measured.

From a scientific perspective, it is not excluded at this time that there could be more than one origin to life. As a result, we often talk about "the origin(s)" instead of "the origin" of life. This subtle, yet fundamental, difference reflects distinct ideas. One proposes that life might have started and gone extinct countless times before taking hold on our planet, long before the first fossils were preserved in the geological record 3.5 billion years ago. This concept is fairly intuitive when thinking of Earth's early environment. The mountain-sized asteroids and comets that regularly collided with the young Earth excavated gigantic impact craters hundreds of kilometers wide and kilometers deep. These impacts were often powerful enough to blast debris back into space and obscure the Earth's atmosphere for hundreds of years or longer. The energy delivered by these violent collisions triggered enormous and long-lasting eruptions. The young Earth itself produced volcanism as it slowly released heat. But another idea proposes that there might have been more than one lineage for life on Earth. While it seems that all terrestrial life goes back to a unique ancestor in what we call the "tree of life," some scientists argue that there could have been more. We will come back to this idea.

WHAT ROLE FOR THE MOON?

Before its crust solidified, our planet did not resemble the world we know today. Instead, it was scorching hot and molten, its surface covered with a magma ocean and isolated islands of solid materials. The discovery of zircons confirms that over 4.4 billion years ago stable landmasses had started to form, and the temperature cooled enough to reach below the boiling point of water. The young Earth was still a violent, unforgiving, barren hell at this stage. And things were about to become a lot worse. Theia, a planetoid the size of Mars, collided with the proto-Earth in what is known as the giant impact hypothesis (or Theia impact). This hypothesis for the formation of the moon is well supported by the characteristics of the lunar rocks and the dynamic of the Earth-moon duo. Before the collision, Theia would have been in a stable position nearly along the orbit of the proto-Earth. Gravitational disturbances from either Jupiter's migration or Venus destabilized it and sent it hurtling toward the Earth.

The giant impact hypothesis has a few alternate versions. In the first one, Theia hit the Earth once in an off-center collision and was destroyed; another proposes a violent head-on collision. The similarity of oxygen isotopes between the Earth and the moon makes this scenario plausible since it would be possible through a high-energy mixing of materials from both protoplanets. In the last scenario, Theia glanced off the Earth a first time, barely escaped destruction, only to come back hundreds of thousands of years later to deliver the final blow. This last hypothesis covers a few unresolved issues with the previous scenarios, including the impact velocity. If Theia had come in too fast, both worlds would have been pulverized; too slow, and the Earth-moon duo dynamics would have been very different. Understanding what happened is critical because this event set the stage for the rest of Earth's evolution. The collision modified the Earth's axial tilt to set it to 23.5 degrees and initiated the seasons of our planet. After the collision, Earth's temperature

The collision between Theia and the proto-Earth played a significant role in our planet's fate, modifying its axial tilt and setting the stage for the seasons as we know them today, which are essential to biological cycles.

rose once again. Any water that would have been at the surface then was vaporized. The remains of Theia were partly incorporated into the Earth, and both cores merged. Vaporized rocks and dust from both Theia and the Earth were blasted into orbit, forming a ring of debris that took merely a hundred years to coalesce and form the moon, maybe less. Right after its formation, the moon would have looked enormous in the Earth's sky, orbiting only 25,000 kilometers away compared to 383,000 kilometers today.

The proto-Earth and moon system did not look anything like it does now, either. The duration of a day would have been just two to three hours long, while the moon orbited around the Earth every five hours. About 1.5 billion years back, the moon was still close enough for a day to last only eighteen hours, and not so long ago, a year was still 420 days instead of the current 365 days. These changes are preserved in 430-million-year-old fossilized corals. Just like trees, corals record their growth in layers. When a coral grows, it accumulates a fine layer of calcium carbonate every day, and its monthly deposits are linked to the lunar cycle. Seasonal changes are visible, too, and in between them, thinner lines indicate daily deposits. This record tells us that during the Silurian period (444 to 419 million years ago), a day would have lasted about twenty hours. A few million years later, other fossilized coral records from the Devonian period (419 to 359 million years ago) show that the Earth's spin had slowed down to 410 days per year. A day then lasted about twenty-one hours. As the Earth continues to slow down, its rotational energy is being transferred to the moon, causing it to pull away from the Earth about three centimeters a year. In time, the length of a day will continue to increase on average by about 0.00001542857 seconds a year to be twenty-five hours long 180 million years from now.[1]

The role played by the cataclysm that formed the moon in the dawn of life, if any, still has to be fully understood. Environmentally, the axial tilt of the Earth imparted by the collision with Theia is responsible for our seasons. The moon's gravitational tug generated tides in the oceans. A twelve-hour day would have produced high tides every six hours. They still reached over three hundred meters high during the Archean eon, 2.5 to 4 billion years ago, pushing the ocean waters kilometers inland. Without them, climate oscillations from the Ice Age to the interglacial periods might have been less extreme than they are. By stirring and mixing cold and warm waters, they drove how much the oceans absorbed sunlight. This stirring is vital, as it produces more predictable and habitable climate conditions and regulates our planet's temperature. Locally, intertidal areas provide intense evolutionary pressures for species living in rapidly changing environments. They may also have contributed to producing the energy necessary to the transition between prebiotic chemistry and life. One thing is certain, however. The formation of the moon certainly played a central role after life took hold, as Earth's seasons produced by the axial tilt directly drive life's rhythms, metabolisms, dispersals, migrations, and speciation over time.

LIFE FROM SPACE?

With this environmental background staged 4.4 billion years ago, science envisions various theories to explain the beginnings of life on Earth. (It also abandoned some along the way, like the Aristotelian notion of a "spontaneous origin of life," where living beings could arise from dust and corpses.) Among the theories and hypotheses still being investigated today is the idea that life did not originate on Earth but came from outer space. This theory is known as panspermia. This term was first found in the antique records of Greek philosopher Anaxagoras (500–428 BCE). Chemist Svante Arrhenius structured it as a scientific hypothesis in 1903.

It has recently found renewed interest in the realization of how ubiqui-tous the building blocks of life actually are in objects of all sizes and types from the solar system to interstellar space. Panspermia conveys the idea that life and the building blocks for it exist throughout the universe and are being delivered by space dust, asteroids, meteors, comets, and planetoids. With the advent of technology, the notion that spacecrafts can inadvertently carry both to another world through contamination was added, and special attention is paid to this question under the label of "planetary protection."

Panspermia provides a mechanism to distribute life and its elemen-tal components throughout the universe readily. It would support the notion that life could be common, as planetary formation by accretion is a universal process. One could thus assume that it only takes a planet to form in the habitable zone of their parent stars and be seeded to have a chance to develop life. The rest would be left to planetary evolution and random catastrophes. We already know of terrestrial microorgan-isms adapted to living in extreme environments. They are known as extremophiles and would make plausible candidate space travelers. Many can endure extreme cold, dehydration, acid, and the vacuum of space, and especially resilient forms could undertake an interplanetary journey given the opportunity. However, the transfer through space is only one part of the panspermia process.

Before microbes can hitchhike an interplanetary ride, they have to survive the ejection from their planet of origin first. Then a transfer in space can last millions of years. If the microorganisms make it that far, the hypervelocity entry into the new host's atmosphere is the next hur-dle before enduring the energy of impact upon landing. Some micro-organisms do better than others in experimental settings, and survival may depend not only on the species but also on the object transport-ing them and the location of the microorganisms on it. For instance, non-photosynthetic organisms sheltered deep within a rock may have a better chance to survive the initial ejection and the atmospheric entry than photosynthetic organisms located much closer to the surface.

Ultimately, it does not matter how much biogenic or biologic material is circulating in outer space if it fails to survive only one of these phases. And, for the lucky ones making it through all these stages, survival in a new world is by no means guaranteed. Therefore, while panspermia may provide a vector for spreading life or its elementary bricks, the ultimate survival and seeding rate might not necessarily be high. But there again, it is a game of numbers. Even if one in a hundred or one in a million attempts makes it through, millions of planets could be harboring life in our galaxy alone. And the more planets are seeded, the more chances they have to become part of the panspermia process by contributing seeds in their neighborhood through planetary exchange, thereby augmenting the probability that life will start somewhere. What panspermia does not do, however, is explain how life begins. It only displaces the question elsewhere.

BIOGEOCHEMICAL THEORIES OF THE ORIGIN OF LIFE

The first modern biochemical model for the origin of life was proposed by Alexander Oparin in 1923 and independently supported by John Haldane in 1928. Their theory suggests that life arose gradually from inorganic molecules, building blocks like amino acids forming first. They then combined to make complex polymers (large molecules composed of many repeating subunits) in warm pools at the water's edge. Oparin did not see any fundamental differences between a living organism and lifeless matter. He thought that the characteristics of life had arisen as part of the evolution of matter. In other words, prebiotic chemistry transitioned into biology. Haldane proposed a very similar idea. For him, the primordial ocean served as a chemical laboratory powered by energy from the sun and lightning. In his model, organic compounds were produced by gases in the early Earth's atmosphere interacting with UV radiation. The ocean became, in Haldane's words, a "primordial soup." This soup was populated by organic monomers and polymers. They would have acquired lipid membranes in time, eventually leading to the first living cells. In 1953, Stanley Miller and Harold Urey tested the Oparin-Haldane model in the lab. They built a closed system that re-created a heated pool of water and a mixture of gas assumed to be representative of the early Earth's atmosphere. They simulated lightning with electrical sparks and let the experiment run for a week.

While no large and complex molecules formed, Miller and Urey demonstrated that eleven of the twenty amino acids used in proteins, sugar, lipids, and other organic molecules could form under these conditions. In 2007, after Miller's death, the original vials from the experiment were reanalyzed with modern chromatographs and mass spectrometers and revealed the presence of twenty-five different amino acids in them. While the composition of the early Earth's atmosphere as simulated by the Miller and Urey experiment has been questioned since, subsequent experiments confirmed that inorganic precursors could produce at least some organic building blocks. Although the exact conditions were still unknown, this was an essential step in understanding the origin of life because it demonstrated that at least some of its building blocks could have formed from non-biological sources on early Earth.

This result was the foundation for the abiogenesis model. It gave us the building blocks of life, but this was not life yet. Oparin thought he was a step closer when his experiments showed the spontaneous formation of small (one- to one-hundred-micrometer) vesicles named coacervates that can provide stable compartmentalization without a membrane. They readily form by the millions upon the cooling of small amounts of hot saturated solutions of proteinoids and can retain their shape for several weeks. Oparin thought they represented an essential stage in precellular evolution. They provide a protective environment for the material inside them, and, from Oparin's perspective, this could have represented an early form of metabolism. However, even if they form spontaneously, these vesicles do not have a mechanism to reproduce, leaving them short of the properties defining living organisms.

According to this biochemical model, the transition from prebiotic chemistry to biology took place in the ocean, where organic compounds aggregated and grew by absorbing nutrients. Like bacteria, they could also divide by budding, a form of asexual reproduction. They would have been intermediate between molecules and organisms at this stage, and some proteins would have acted as enzymes and began metabolic activities. To summarize the process described by Oparin and Haldane, it all started with free atoms, then simple inorganic molecules, followed first by simple organic molecules and simple organic compounds that then evolved into complex organic compounds like the ribonucleic acid (RNA), the deoxyribonucleic acid (DNA) and proteins, and then the first forms of life.

The first organisms that developed from these conditions were heterotrophs, which means they could not produce their own food. Instead, they derived their energy from the fermentation of the ocean's primordial soup. They also did not need oxygen to survive, which defined them as anaerobic. Autotrophs, organisms that can produce their own food in the form of complex organic compounds such as carbohydrates, fats, and proteins, using carbon from simple elements like carbon dioxide, appeared 2 billion years ago. They took their energy from sunlight through photosynthesis or inorganic chemical reactions (chemosynthesis).

Oxygen evolved from photosynthesis and built up in the Earth's atmosphere, transforming it into an oxidizing environment. From there on, life would evolve to adapt to oxygen and aerobic respiration. To some extent, this period can also be seen as the beginning of the first form of a free-market economy, with food producers and food consumers.

Another biochemical theory, the "RNA world," simplifies the process leading to the origin of life. DNA, RNA, and proteins play a critical role in forming life. DNA can store genetic information, and proteins can catalyze the reactions. But RNA has the advantage that it can do both without any help. It has catalytic power and can self-replicate, allowing it to propagate its chemical identity over generations. This particular theory thus proposes that life began with simple RNA molecules that could copy themselves without help from any other molecule. An experiment performed in 2022 by researchers at the University of Tokyo appears to support this theory.[2] For the first time, they were able to create an RNA molecule that replicates, diversifies, and develops complexity following Darwinian evolution, which means that it could evolve and adapt over time. Their experiment provided the first empirical evidence that simple molecules can lead to the emergence of complex, replicating natural systems.

CONDITIONS FOR LIFE'S INCEPTION

The environmental conditions on the young Earth gave the physical and chemical constraints for prebiotic chemistry and early life. And they, too, are subject to heated debates. For example, the Miller-Urey experiment was based on Darwin's concept that molecules underwent reactions in warm little ponds that led to the formation of prebiotic compounds like amino acids. With time and further reactions, these prebiotic compounds would have produced increasingly more complex molecules, ultimately leading to life, but others argue that life's incubator was instead a very hot one, which would make sense knowing the Earth's environment during the Hadean eon. In order to replicate materials from volcanic eruptions, experiments added hydrogen sulfide and sulfur dioxide to Miller's mix, which, indeed, resulted in a more diverse set of organic compounds. The discovery of Archean bacteria surviving in hot springs at temperatures as high as 92 degrees Celsius could bring additional support to the hypothesis of a hot origin for life. Chemical reactions also occur faster at high temperatures, which would be another benefit of a hot origin.

But not so fast! Organic compounds are actually more stable at cold temperatures. Different types of organisms living in extreme, and this time, very cold environments can still metabolize with temperatures around minus 20 degrees Celsius. Could life have started on a cooler Earth instead? One reason to seriously consider a cold origin of life is that the sun was much less luminous earlier in our planet's history. Less energy received by the surface could have meant that at least some of the Earth's oceans were covered in ice, possibly hundreds of meters of it. In the context of the RNA world hypothesis, these conditions would have provided stability to the RNA molecules and confinement to react and replicate, which has been successfully tested by laboratory experiments.

Other environments have drawn the attention of scientists over the years, two in particular, which are still strong contenders and continue to trigger passionate discussions. One is the deep-sea hydrothermal vents hypothesis, and the other considers hot springs on land. The first one proposes that deep-sea hydrothermal vents provided ideal conditions for the origin of life, as seawater comes into contact with minerals from the Earth's crust and reacts to create a warm, alkaline environment containing hydrogen. Mineral-rich chimneys form in the process. In them, both alkaline and acidic fluids circulate and produce the energy necessary for chemical reactions between hydrogen and carbon, helping the formation of increasingly complex organic compounds. A team led by Nick Lane at the University College London successfully generated protocells in hot alkaline water in 2019, in a breakthrough consistent with the idea that this type of environment could have been where life originated.

Shortly after this result was published, Bruce Damer and David Deamer at the University of California Santa Cruz provided supporting evidence for an origin of life on land rather than in the ocean, this time in fluctuating volcanic hot spring pools.[3] These new experiments showed that lipid-encapsulated polymers could be synthesized by cycles of hydration and dehydration to form protocells.[4] The cycling through wet, dry, and moist phases appears essential to boost combinatorial selection, including primitive metabolic activity. Such an environment could have led to the generation of progenotes, an organizational structure that preceded single-celled organisms. After undergoing selection and distribution, these progenote populations dispersed and adapted to new environmental niches, establishing networks that ultimately led to the first microbial communities. This theory could create a transitional pathway between prebiotic chemistry and the emergence of life that would start with the formation of simple protocell aggregates, followed by the transitional form of progenotes, then the production of robust microbial mats formed by single-celled prokaryotes that left the first fossils in the Archean geological record.

John Bernal presented a very different approach in a book titled *The Physical Basis of Life* (1951), with the possible role of clay minerals in the origins of life. Clay minerals would have been readily available in the early Earth environment, formed by the weathering of volcanic glass and rocks by water. The crystal lattice of clay mineral particles could have helped concentrate organic compounds together and organize them into patterns, much like our genes. Clays could have also been critical in the chemical evolution and the origins of life because of their ability to take up, concentrate, and catalyze the polymerization of organic molecules, and protect them from UV radiation. In 1982, A. Graham Cairns-Smith suggested that clays could store and replicate structural defects, dislocations, and ionic substitutions and act as "genetic candidates." In short: clays are catalysts and can replicate. Some argue that they could have evolved up to metabolism. For them, life began in the clay world and not in the RNA world.

These theories are still evolving today as our knowledge of early Earth's environment advances and our understanding of primordial conditions improves. And the more they do, the more exciting the unfolding story becomes. In 2012, Fred Ciesla and a NASA Ames colleague of mine, Scott Sandford, simulated the conditions of the solar nebula 4.5

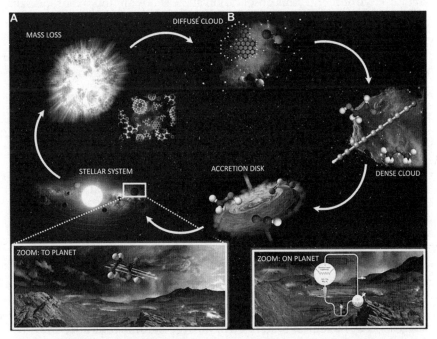

The life cycle of a star and the corresponding molecular complexity. The insets (left) show how molecules are delivered on a planet, for example by comets and asteroids, and (right) how they then evolve depending on the environment.

THE SECRET LIFE OF THE UNIVERSE

billion years ago, at a time when planets were forming. Their experiment demonstrated that organic compounds are natural by-products of the evolution of the protoplanetary disk (the circular cloud of gas, dust, and ice from which planetary systems are formed) and that they are found everywhere. The implications of this experiment are profound: proto-Earth came preloaded with the building blocks of life.

NEW CONCEPTS

Today, we still do not know what exact environmental conditions led to life, but here is what we know for certain: the building blocks of life are everywhere. They seem to be the easiest things to form in a planetary nebula environment. This observation alone sends an important message for the search for life beyond Earth. But there is another equally powerful message embedded in this result. When it comes to the environments in which life could have originated, we can only test various hypotheses to the best of our current abilities: warm, hot, cold, acidic, alkaline, and anywhere in between, and see how the chemistry works out. But there, too, a transition from prebiotic chemistry to life appears possible in more than just one scenario, as demonstrated by the various theories. The building blocks of life are easy to form, and theoretical and empirical approaches show that they can assemble in a multitude of extremely different environments. This observation has led some to wonder whether life is inevitable, evolving out of necessity rather than by accident, simply by following nature's physical and chemical laws.

This "inevitable life theory," proposed by Jeremy England in 2013,[5] suggests that biological systems will emerge naturally because they dissipate energy more efficiently. In other words, life could be an inevitable product of thermodynamics. To some extent, this resonates with other recent theories advocating for multiple origins of life realized through multiple pathways, such as proposed by Chris Kempes and David Krakauer of the Santa Fe Institute.[6] These concepts are probably the closest to a new intellectual framework we have today.

Existing models generally consider life to be the result of a successive chain of evolution that started with its origin and continued through cycles of extinctions and adaptation. Kempes and Krakauer approach the origin of life differently, in the same way we think of converging (or analogous) evolution for some of nature's inventions. For instance, the eye was not invented once by nature, but multiple times and independently. Likewise, they propose that life could have emerged more than once independently. They consider first all the combinatorial possibilities from the materials that could form life, then the constraints that limit its potential, and the optimization processes that drive adaptation.

In their most recent models, life is viewed as adaptive information. They use the analogy of computation to represent the processes central to life, with some interesting outcomes. For instance, not only could life have emerged multiple times independently, but some apparent adaptations could be new forms of life instead. Another intriguing outcome is to make life a continuum and allow for an actual (and much broader) definition of life, including, as Kempes mentioned while referring to his work, "cultures that live on the material of minds in the same way that multicellular organisms live on the material of single celled organisms." Instead of being "something," life is a process, an approach that begins to touch on the nature of life. This model provides space for what is known as the shadow biosphere,[7,8] a hypothetical biosphere on our planet based on entirely different molecular processes and biochemistry than those of the life we know. Its life-forms would be so different that we would not be able to recognize them as living. It is not always that clear-cut in our biosphere in the first place: discoveries shake our confidence from time to time, blurring the line between living and non-living. For instance, some giant viruses have complex genomes, even though they rely on their hosts for essential biological functions. In other words, we should expect the unexpected.

CHANGING PARADIGM

There are potential paradigm-shifting discoveries that could completely revolutionize our approach to the origin of life in the near future. For example, suppose life arose more than once on Earth. But we are limited in what we can see because when we analyze samples, we recognize what is familiar to us. Given our limitations, as stated by Carol Cleland about her work, "shadow (nonstandard) microorganisms might never turn up in our tests because they are designed for biology as we know it." But on the other hand, nonstandard organisms most likely would also leave traces of their interactions with the environment, the kind we cannot easily explain because maybe they don't quite fit our models and predictions. That is why I devote most of my research with my team to understanding this interface between life and environment, that nexus where life leaves its fingerprints on its habitats, because the most extraordinary and transformative discoveries may be awaiting us there.

The multiple pathways to the multiple life hypothesis and the shadow biosphere can also lean on the fact that the building blocks of life on Earth are like letters of the alphabet. The same letters in a different order can make different words with meaning. They can also be garbled. Other letters may lead to different words, but ones that have a close meaning (synonyms). The question now is whether different molecules can do what our DNA does, which is to carry the genetic instructions for the development, functioning, growth, and reproduction of living organisms. Maybe different alphabetical combinations have developed on early Earth and are still around, but we cannot recognize them. And DNA is not the only place where life on Earth could have diverged.

"Standard" terrestrial life is based on twenty amino acids. We already know that a hundred of them exist and naturally assemble in the lab. The possibilities are thus almost limitless, especially since each amino acid can have different geometries. Those we make in the lab result in two mirror images of each other, called left-handed and right-handed, but the great majority of them are left-handed in nature, which means that molecules twist in the same direction. This natural shape selection is called chirality, and we are not sure why life on Earth is primarily left-handed. Amino acids discovered in the Murchison meteorite in Australia might contain a clue. They were found to be, for most, left-handed, too, possibly an indication that DNA building blocks may have taken shape in deep space. So, could a shadow biosphere be "ambidextrous" or a mirror image of ours? More important, how do we test the shadow biosphere hypothesis?

Whether it exists or not, the pursuit of a shadow life on our planet will bring the intellectual and technological advances that may put us on track to discover life beyond Earth because it teaches us what it takes to recognize life as we don't know it. And, on that count, we have a lot to be looking forward to, since the vast majority of terrestrial species (mostly microbes) and their relationship to the environment remain to be characterized today. Some might reveal the multiple origins of life on Earth. More to come here, as that scientific chapter still has to be written.

Meanwhile, even if we could verify the existence of a shadow biosphere on Earth, we would still be a few steps away from understanding how prebiotic chemistry became biology, be it standard or weird. Our chances to find out the answer on our planet are slim, except maybe coming as a result of a lab experiment one day, in a modern version of the Miller-Urey experiment. Unfortunately, the same dynamics that make our planet so unique in the solar system are also what may likely prevent us from discovering our origins here on Earth. This record is gone—forever lost to geological recycling. Plate tectonics and erosion erased nearly everything older than 4 billion years. When it is not gone, like the 4.28-billion-year-old Nuvvuagittuq greenstone belt in Canada, the remnants of these ancestral rock beds are too few and far between to give us a good statistical chance of making a discovery showing us how prebiotic chemistry became biology. But, there is still a possibility that we could access that evidence one day and in the most unlikely places of all: at the edge of the habitable zone of our solar system.

3

VENUS AND ITS
VEIL OF SECRECY

When we look at the solar system's habitable zone, where water can remain liquid at the surface, three of the four inner planets immediately stand out: Venus, Mars, and Earth. Nearly 108 million kilometers away from our star, Venus sits within the inner (hot) edge, while Mars is near the outer (cold) boundary 240 million kilometers from the sun. Venus, Mars, and Earth in the middle: these worlds give us a unique opportunity to observe how fragile habitability can be and how environmental change can rapidly turn a habitable world into a planet hostile to life.

Earth was once thought to be resting comfortably within the solar system's habitable zone, but new studies have refined those calculations, showing that our planet could be closer to the inner edge than previously thought, and much too close for comfort, as it makes it easier to push it over the limit. The Earth is fine, for now. But the inner solar system was better off early in its history, when these three worlds were much more similar than they are today.

At one time, the solar system harbored not one but potentially three habitable planets. Venus, Earth, and Mars had substantial gravity and atmospheres that were similar in composition. External and internal

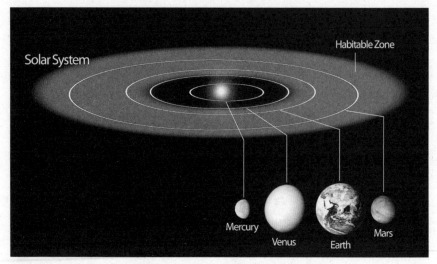

Habitable zone of the solar system, with Venus and Mars at both ends.

geological processes shaped all three, and they developed an intense volcanic activity that added volatiles and carbon dioxide to their atmosphere. It is also possible that all three had oceans. But from their early start, these three worlds were bound to experience very different evolutionary trajectories. As of today, only the Earth has a known biosphere. We are currently exploring the possibility that life developed at some point on Mars. If it exists, hostile conditions on its surface for the past 3.5 billion years would have rapidly confined it to the subsurface and deep underground. Still, Mars remains a high-priority candidate for a biosphere and the focus of an integrated program of biosignature exploration. Things are more complicated for Venus.

THE PHOSPHINE MYSTERY

A recent study made headlines suggesting that the spectral signature of phosphine had been discovered in the atmosphere of Venus.[1] Phosphine

(PH_3) is a toxic compound of hydrogen and phosphorus. It is only produced by life on Earth, be it human industry or anaerobic bacteria on
Earth, which are microbes that cannot live or grow when oxygen is
present. The discovery was exciting because of where phosphine was
allegedly detected. Sixty kilometers high in the Venusian atmosphere,
pressure and temperatures are within range of those on Earth, thus far
away from the hellish conditions of the surface, where the temperature
is 465 degrees Celsius and the pressure ninety-three times that of Earth.

How microbes produce phosphine is uncertain. In short, phosphine is
found where microbes are, but we still have to understand how they produce it. On Earth, one way would be for them to take up phosphate from
minerals or biological material, add hydrogen from water in their environment, and ultimately expel phosphine. On Venus, the gas should typically
be broken down by a highly acidic atmosphere, and there has to be a continuous source somewhere to replenish it, if it is present. Such a source can
either result from biological activity or a process we have yet to recognize.

How living organisms may produce phosphine in the clouds of
Venus is only one of several issues to deal with. Another relates to the
observation itself, since only one absorption line was identified in the
spectral data, putting the claim on shaky ground from the beginning.
As a result, teams of scientists decided to undertake follow-up research
to verify the observation. A study reexamined the data from one of the
telescopes used to make the claim and concluded that it could not find
the spectral signature of the gas. The detection process can be tricky
for many reasons. Some relate to the nature of sometimes weak spectral
signatures or their proximity to one another, which may lead to confusion. Others may result from issues with data processing or analytical
tools that can lead to erroneous conclusions. Having only one spectral
line to work with did not make the interpretation any easier. Another
verification study used modeling to calculate how gases would behave in

the Venusian atmosphere. Its results suggest that instead of phosphine, the gas might actually be sulfur dioxide, which, contrary to phosphine, is common in the atmosphere of Venus, and it would be present more than eighty kilometers above the surface, not fifty to sixty kilometers up, as initially proposed.

Yet, those who led the verification studies are the first to admit that these results cannot completely rule out the phosphine hypothesis, as data from NASA's 1978 Pioneer Venus mission could come to support the original observation. As it descended through the atmosphere, Pioneer Venus dropped a probe to measure the cloud chemistry and detected phosphorus. But whether the signature was phosphine or another phosphorus compound is unknown. Ultimately, only a mission to the clouds of Venus to analyze these gases will bring closure, and this could happen soon. NASA, the European Space Agency (ESA), and Rocket Lab, a private company, have a flotilla of planned missions to Venus that will be launched within this decade and the beginning of the next. In the meantime, let's step back for a moment and assume that, despite skepticism, the phosphine signature turns out to be confirmed. After all, phosphine on Earth is a biosignature, whether it is the result of human industry or microbial activity. So, what are the odds of it being produced by microbes living and drifting in the clouds of Venus?

Although I did not contribute to the verification studies, I found it an excellent time to remind myself of Carl Sagan's ECREE standard (extraordinary claims require extraordinary evidence) and look at where the body of evidence might take us. I was not the only one thinking along the same lines. In a study proposing volcanic instead of biologic processes for phosphine,[2] Ngoc Truong and Jonathan Lunine paraphrased Sagan's ECREE standard, too, by saying that "the hypothesis of life producing phosphine in the clouds of Venus requires both the extraordinary claim that life exists in the clouds and a mechanism to maintain its viability."

Although planetary and environmental context might not be everything, they give us, more often than not, a solid foundation to shed light on where new observations fit. They also help us put these new data into the perspective of an existing body of knowledge. But let's assume for a moment that there is life on Venus and focus on the process it would take to maintain it in the atmosphere using Earth as an analogy. On our planet, the atmosphere is neither an environment where life originates nor can it be sustained. Clouds are too unstable to allow life to develop in them and maintain a habitat. On Earth, the atmosphere simply acts as a transportation agent. Bioaerosols, such as bacteria, fungi, algae, and pollen, are present everywhere in it. Their types and concentration may vary significantly in different locations and under various atmospheric conditions, but they are only there in transit.

Microorganisms are originally sourced on land and over the surface of the oceans. They are then carried away in clouds and plumes of dust by wind turbulences or volcanic eruptions and may also act as cloud condensation and ice nuclei. As they do, they influence the hydrological cycle and the climate and, in many ways, facilitate the mechanism (precipitation) that brings them back to the surface. But it can be argued that, while Venus and the Earth share some similarities, they are still two very different worlds, starting with their atmospheres. Would it be possible, then, that a biological hypothesis for phosphine could find support in those differences? Venus's atmosphere is dominated by carbon dioxide (96 percent) and nitrogen (3.5 percent). In comparison, the Earth's atmosphere is composed of nitrogen (78 percent) and oxygen (21 percent). Venus has a large convective region (the troposphere) that extends up to sixty kilometers above the surface, about four times higher than the Earth's troposphere. Between sixty and eighty kilometers altitude, in the region where phosphine might have been detected, permanent opaque clouds made of sulfuric acid droplets obscure the surface entirely. There, pressure and temperatures are close to Earth's. This thick atmosphere generates the strongest runaway greenhouse effect in the solar system. As a result, surface temperatures are hot enough to melt lead and boil out most water vapor to the point that the effective concentration of water molecules is about a hundred times too low for even the heartiest known organisms to survive. Since water is biochemically essential to life as we know it, this could be considered a fatal blow to any idea of biology developing in the clouds.

Venus captured by the ultraviolet imager of the Akatsuki probe on December 23, 2016. The image shows the sharp contrasts between the tropical regions covered with convective clouds and the calmer polar regions.

But Jane Greaves, the lead author of the phosphine study, does not give up just yet. For her, as long as we do not know how well the atmosphere mixes, a livable window can exist in the Venusian clouds. In it, some cloud droplets could have very high water content. Could those explain the location of dark streaks sometimes observed on short timescales in ultraviolet light in that unique region of the atmosphere? Could those be microbial colonies that evolve, die out, and reemerge? At this point, it is only pure speculation.

We also need to examine what could be the source for these hypothetical living organisms. The idea that microorganisms originate at the surface like on Earth faces serious challenges. Venus's surface conditions are so extreme that it is hard to imagine anything surviving on the ground. Further, while winds around 360 kilometers per hour keep the clouds in constant motion—circling the planet every four days, their speed drops considerably near the surface, where they only reach a few kilometers per hour. They are simply not strong enough to scoop up any material and transport it sixty kilometers high. Unfortunately, as much as we all would like to embrace the idea that Venus is inhabited, conditions appear beyond the range of what life (as we understand it) can tolerate. Alternatively, suppose we let the data guide us. Then phosphine could fit very well in the context of the current Venusian environment and may actually point toward a very compelling discovery.

AN ACTIVE VOLCANISM ON VENUS?

As part of the verification work, researchers also looked into hypotheses that do not involve life. For example, trace amounts of phosphides formed in the mantle of Venus are a credible non-biological source for Venusian phosphine. In this hypothesis, small amounts of phosphides originating from the deep mantle are brought to the surface by volcanism. Explosive eruptions provide enough energy to launch them kilometers high into the atmosphere as volcanic dust, explaining the episodic changes in sulfur dioxide observed in the atmosphere. Once there, the phosphides would react with sulfuric acid to form phosphine. Assuming this is correct, phosphine may tell us something profound about the nature of the Venusian mantle and, more excitingly, about volcanoes still active on Venus today.

We have known for over a quarter of a century that Venus is a volcanic world but the atmosphere is so thick and hazy that it is impossible to tell if volcanism is active. We can only see the surface using radar, which is how, in the early 1990s, NASA's Magellan spacecraft revealed a planet covered in 1,600 major volcanoes and possibly one hundred thousand to a million more minor volcanic structures and extensive lava flows. Fifteen years later, ESA's Venus Express mission measured the amount of infrared light emitted from parts of the surface. These data were combined with laboratory experiments that re-created Venus's atmosphere and its interaction with minerals abundant in basaltic lava flows. The results showed that infrared radiation from three volcanic regions differed from their surroundings and their surface was much less weathered.

A few years later, a sharp rise in sulfur dioxide in the upper atmosphere was followed by its gradual fall and interpreted as a possible sign of volcanic activity. More recently, thermal mapping detected localized changes in surface brightness between images only a few days apart. If the phosphine signal is real on Venus, it would be consistent with these observations and what we know of the environment and geology. But, of course, that does not make it irrefutable proof that volcanoes are currently erupting, yet it would fit nicely with that hypothesis. It would also add to a mounting and converging body of evidence, which contrasts with an elusive biosignature for which we have to decide how much we want to bend backward to make the data fit the conclusion. Ultimately, only measurements in the atmosphere itself will bring a definitive answer. Meanwhile, beyond all the controversy, the phosphine claim raises exciting and meaningful questions, not the least of which is that we do not yet know any unambiguous signature of life. But the question of Venus's habitability and life potential goes beyond the phosphine debate and started well before it, and for good reasons, because Venus is a fascinating world!

A VERY STRANGE PLANET

The Earth and Venus probably had comparable environments in the early days of the solar system. However, except for a very similar size and density, Venus does not have much left in common with our planet today. In fact, it is one of the oddballs of the solar system. Its axial tilt is almost perpendicular to the orbital plane, and it spins in the opposite direction from most other planets. The sun rises in the west and sets in the east. Stranger yet, the planet slowly rotates around its axis, but goes around the sun relatively fast. As a result, a Venusian day, which equals 243 Earth days, lasts longer than a Venusian year, which equals 224 Earth days. Stated differently, one would say happy birthday on Venus more often than good morning!

Further, although flipped on its axis, Venus has a minimal inclination angle and a nearly circular orbit around the sun, so it has no seasons, and temperatures remain stable throughout the year. It is steadily hotter

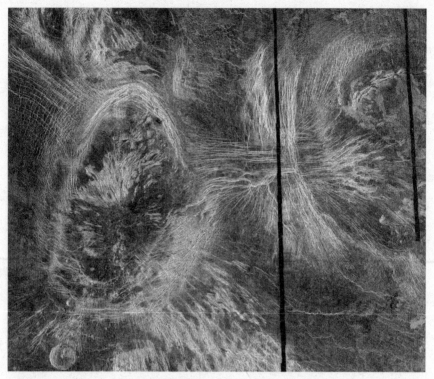

Bahet (230 by 150 kilometers) and Onatah (350 kilometers in diameter) are two coronae on Venus. These crown-like volcanic features, seen here in a mosaic from NASA's Magellan mission, are surrounded by ridges and ditches, with domes in their centers. Their formation is likely due to rising magma. A pancake-shaped volcanic structure is also visible on the bottom left.

than it is on the surface of Mercury, although Venus is farther away from the sun. This hellish temperature results from a catastrophic greenhouse effect in the carbon dioxide atmosphere. Sunlight is trapped by the atmosphere and cannot escape into space, which boosts the surface temperatures by about 390 degrees Celsius compared to what they should be. Venus would still be hot without this greenhouse effect, but it would be a very different world. Whatever triggered the runaway greenhouse effect was what ultimately doomed the potential for life on Venus, and here, there are various schools of thought.

DIFFERENT SCENARIOS FOR
HABITABILITY ON VENUS

Early similarities with the Earth make it possible that Venus had liquid water and maybe oceans on its surface. This hypothesis is consistent with the deuterium-to-hydrogen ratio inferred from NASA's Pioneer Venus mission data. Deuterium (a hydrogen isotope) and hydrogen escape into space at different rates. Their ratio allows the measurement of how much is missing from the original reservoir. How much water was outgassed (and when) remains an outstanding question. Four billion years ago, when the young sun was only 75 percent as bright as now, Venus would have been comfortably located within the habitable zone. But as the sun became more luminous, the planet was pushed closer to the hot edge. With the temperature rising, oceans would have begun to evaporate, releasing vast quantities of water vapor into the atmosphere. Water vapor is a greenhouse gas; thus, more water absorbs more heat, which increases the temperature even more. A positive feedback loop worsens evaporation until there is nothing left to evaporate.

And there are more nefarious effects to this feedback mechanism. Water is a critical lubricant in the plate tectonics process. Without oceans, plate activity stops, and, with it, a key regulator of the amount of atmospheric carbon dioxide disappears. Instead of being recycled at depth by the movement of plates, carbon on Venus ended up trapped close to the surface. It was subsequently injected into the atmosphere through volcanic eruptions and slow outgassing. Making matters worse, Venus does not have a magnetic field, which allows the electric field carried by the solar wind to penetrate the planet's ionosphere, stripping away its ions, including oxygen and hydrogen. In this scenario, Venus would have stopped being habitable for life as we know it between 1 and 2 billion years after its formation.

In another scenario, like the Earth, Venus had a magma ocean for about a hundred million years after its formation and a carbon dioxide atmosphere. At this stage, any water would have been present only as steam. Running the clock forward, and Venus being closer to the sun than the Earth, the temperatures never fell low enough to allow the formation of raindrops and precipitation. Instead, the water remained in its gaseous form, preventing oceans from ever forming. Clouds tended to develop more at night, and their formation never allowed the planet to cool rapidly enough. In that regard, a less luminous young sun may have saved the Earth from the same fate. Weaker solar radiation on a young Earth located just a little farther away than Venus from the sun might precisely have been what allowed the formation of the terrestrial oceans. This scenario could also provide an unexpected resolution to the "faint young sun paradox," an issue that has vexed the scientific community for a long time. Instead of hindering the development of life on a cold Earth, a weak young sun would have supported it by keeping the temperature within a range that prevented a runaway greenhouse effect on our planet. It would have allowed evaporation, condensation, and precipitation, and the formation of Earth's oceans.

In the third scenario for habitability, the loss of oceans would have taken place only recently. It would be due to the simultaneous eruption of large igneous provinces that ended Venus's temperate period. This scenario is the most favorable since it predicts that Venus was habitable for most of its history until only a few hundred million years ago. These eruptions could explain the very young surface of Venus, which is only between 300 to 700 million years old, and the runaway greenhouse effect through the release of large amounts of carbon dioxide in the atmosphere. During the same period, newly exposed basalt became efficient oxygen sinks.

These represent vastly different visions for Venus's habitability over time. One extreme suggests that Venus was never habitable, while another proposes that it remained habitable until recently at a geological scale. Figuring out which scenario is the most likely will require a better understanding of how hydrogen escaped into space over time. The newly planned missions should help us evaluate this question with greater accuracy. Eventually, even in the best-case scenario, habitable conditions on Venus ended a few hundred million years ago with the synchronous eruption of all the large igneous provinces. They never erupted all at once on Earth, which possibly is why we are here today. Such eruptions are catastrophic extinction-level events on a planetary scale. If life were on Venus then, it would have had very little time to adapt to such a profound environmental change. In particular, considering the phosphine hypothesis, life would have needed to make a rapid and drastic change of habitat from the surface to the atmosphere. Moreover, the stability of the cloud cover on Venus, often mentioned in support of the possibility that life developed a habitat in the atmosphere, might not be such good news, either. Life needs some dynamic change to evolve and boost speciation.

A SUPERCRITICAL FLUID

And so, for the past few hundred million years, the only "ocean" left on Venus has been one of a very different kind. Conditions are so hostile that the combined temperature and pressure are high enough at the surface to turn carbon dioxide into a supercritical fluid, a point where liquid and gas phases do not exist anymore. A terrestrial equivalent would be water coming out of black smokers, those hydrothermal vents forming under tremendous pressure kilometers deep in the ocean. As a result, the

Venusian atmosphere at ground level may feel like a sea of sorts covering the entire surface. It transfers heat very efficiently, buffering temperature changes between night and day and keeping the planet hellishly hot at all times.

Assuming life ever developed on Venus and survives somehow in these conditions, its biochemistry must be drastically different from Earth's. It would have to thrive in an extremely dry environment without significant variability to support evolution and speciation. This scenario is almost the perfect opposite description of Earth's biological processes, all bathed in a cauldron of sulfuric acid much nastier than the deadliest terrestrial hot spring. But if life is there, new missions will make us discover something incredible, something beyond our wildest imagination: life as we cannot fathom it. And, if life never appeared on Venus or did not survive, we will have finally found an environmental limit to it, which is so elusive here on Earth. And we might be about to find out.

PROMISING MISSIONS

As it turns out, the first week of June 2021 was an auspicious one for the exploration of Earth's sister planet. Between June 2 and 8, 2021, new missions aiming for Venus were selected by NASA as part of its Discovery Program (VERITAS and DAVINCI) and by ESA (EnVision), a medium-class mission in the ESA's Cosmic Vision plan. These three missions will work in concert to deliver the most comprehensive study of Venus yet. While their objectives are specific, with some overlaps, a common theme is to understand why Venus and Earth evolved so differently and what we can learn from it for habitability and life.

VERITAS originally planned to launch in 2027 but now might be delayed to 2031. Its radar will see through the clouds to map the surface. In addition, the spacecraft will collect data on Venus's gravitational field to determine the planet's interior. VERITAS's goal will be to bring an understanding of the planet's past and present geological processes from its core to the surface. For instance, the Earth and Venus have developed very different tectonic processes, which might help us test hypotheses on how continents were formed on Earth. Volcanism will also be a primary focus of the mission, including the composition of volcanic deposits and the possibility of ongoing eruptions. VERITAS will investigate the role of tectonic and volcanic activity in the evolution of Venus's habitability, particularly the effect they had on the evolution of the atmosphere and climate. From the comparison with the Earth, the mission might help us understand better what makes a rocky planet habitable and why life may or may not emerge.[3]

DAVINCI (Deep Atmosphere Venus Investigation of Noble gases, Chemistry, and Imaging) should launch in 2029 and focus on the evolution of Venus's atmosphere and its differences from those of Earth and Mars.[4] It will also investigate the possibility of an ocean in the past and the evolution of the lower atmosphere. In complement to VERITAS, it will study the tectonic processes to understand how terrestrial planets form. The mission will be the first time in almost fifty years that a probe descends to analyze the Venusian atmosphere and return images from the surface since the Soviet Venera 13 in 1981.

The European Space Agency's EnVision[5] should launch in 2031, and NASA will collaborate with this mission. EnVision aims, in part, at understanding whether Venus was habitable and had an ocean and how the planet released its heat throughout its history. Its investigations include the chronology of geological events at the planet's surface and a characterization of its morphological diversity. A spectroscopic suite (VenSpec) will collect compositional data analysis of rocks, perform high-resolution atmospheric measurements, monitor sulfur gas to track potential volcanic eruptions, and identify the nature of the mysterious dark patches in the clouds of Venus's upper atmosphere that absorb UV light.

But a private company might reach Venus first with the Venus Life Finder (VLF) mission.[6] VLF represents a series of focused astrobiology mission concepts to document habitability and search for the chemistry of life in Venus's atmosphere. This series of missions will not be affiliated with NASA, ESA, or SpaceX, but with the Rocket Lab company, which plans to launch its own rocket as early as December 2024. The spacecraft will use lasers to search for complex organic chemistry and life inside the clouds of Venus and will test the phosphine hypothesis. Once at Venus, the probe will skim the clouds for about three minutes, but the VLF team intends to make repeated visits to Venus over several years.

This wave of spacecrafts in the upcoming decade should finally lift Venus's veil of secrecy. Their overlap is an added guarantee that, should one fail, another mission could recover the data. So much is to be learned about Earth's sister planet, and a return was overdue. By being just barely closer to the sun than the Earth, Venus already experienced what will be the fate of Earth a billion years from now, or earlier, if human activity continues to alter its environment. In more than one way, its exploration will teach us what impact climate change can have on habitability and life at a planetary scale and maybe how to navigate it.

4

BLUE SUNSETS

On the opposite side of the habitable zone from Venus stands Mars, a high-priority target for the search for life in the solar system and an unwavering focus for human imagination. From the tales of war gods to modern science fiction, Mars has populated the psyche with dreams of neighboring alien life for millennia. First observed with a telescope by Galileo in 1610, the red planet was perceived by Giacomo Miraldi to have "white spots" in 1704, which he later speculated could be ice caps. Astronomer William Herschel was the first to note a faint atmosphere in 1784 and mistakenly interpreted the dark areas on Mars as oceans and the brighter regions as land. He also measured the 25.19-degree axial tilt of the planet, which is close to Earth's 23.5 degrees. However, polar cap deposits observed from orbital spacecrafts showed that Mars's axial tilt has much wider variations than Earth's and oscillated between 15 and 35 degrees in the past five million years, while the Earth's only varied by 1.4 degrees. These large swings produce more frequent and marked climate changes on Mars. Since the planet orbits farther away from the sun and takes twice as long to go around it (687 days), Mars's axial tilt generates four seasons that last longer than on Earth. During the year, temperature extremes vary between +20°C at the equator and –125°C at the south pole. They can also fluctuate extremely rapidly with minor changes in topography due to an atmosphere 165 times thinner than on our planet.

The two small moons, Phobos and Deimos, were discovered within a week of each other in August 1877 by astronomer Asaph Hall, who also determined the mass of Mars. In the late nineteenth century, imagination was sparked by Giovanni Schiaparelli and Percival Lowell's telescopic observations of what they believed were oases and canals. Today, these so-called canals have taken their place in the gallery of ancient myths. We now understand them to be the product of an optical illusion caused by the chance alignment of craters and dark features viewed through telescopes.

MARS: A MYTH ALWAYS RENEWED

Many were still convinced that Mars would be covered in crystal domes and futuristic cities when Mariner 4 left on November 28, 1964, for what would become the first successful flyby of the red planet. Ground-based observations gave additional reasons to hope.

Expectations were thus high that life, possibly even a technologically advanced civilization, had evolved in parallel with us just a little farther away from the sun. And then, on July 14, 1965, after an eight-month journey, it only took twenty-one images to annihilate millennia of dreams and fantasies. Mariner 4 flew 9,800 kilometers away from the red planet at its closest approach, and the images it took, black and white and grainy in quality, covered about 1 percent of the entire surface. In the images, Mars did not look that much different from the moon. So the myth of an advanced civilization had to be abandoned.

But Mars was about to recapture popular imagination with Mariner

This image of the Martian surface, taken by Mariner 4 on July 15, 1965, from a distance of about 12,500 kilometers, reveals a lunar-like surface with an impact crater about 150 kilometers in diameter. Initial disappointment was quickly erased by the discoveries of the following missions.

9, and it has never let go since. On November 14, 1971, Mariner 9 became the first spacecraft to orbit another planet and collect data about the Martian atmosphere and its moons. Surface imaging was delayed by a dust storm that became one of the largest global storms ever seen on Mars, and completely obscured the surface for three months.

Once the atmosphere cleared, the spacecraft began its observation campaign, capturing 7,329 images that depicted over 80 percent of the planet. It also collected data on the atmosphere and the surface. This mapping resulted in the first detailed views of Valles Marineris, the largest canyon in the solar system, Tharsis's giant shield volcanoes, the polar caps, and the moons. Although the resolution was still low by today's standards, Mariner 9 was a paradigm-shifting mission. The red planet had evidently experienced a very complex geological history and had enough similarities with early Earth to warrant further investigation on the possibility that prebiotic chemistry and microbial life might have emerged there, too. The search for life became one of the objectives of the mission that came next: the Viking mission.

DID VIKING FIND LIFE ON MARS?

Viking 1 and 2 were launched within three weeks of each other on August 20, 1975, and September 9, 1975, each orbiter carrying a lander. Viking 1 successfully landed on July 20, 1976, and operated in Chryse Planitia for over six years, well beyond its designed lifetime of ninety days. Viking 2 followed on August 7, 1975, and landed in Utopia Planitia, where it remained operational for three and a half years. Combined, both orbiters and landers would deliver 52,663 images, covering 97 percent of the planet. In addition, the landers themselves captured over 4,500 close-up images of the surface.

The data from the Viking mission remained the foundation of our knowledge of Mars for twenty years and an extraordinary database to investigate. I remember preparing both my master's and my PhD theses with them. The images were still printed on photo-quality paper, filling metal cabinets in my office at Meudon. When I defended my PhD thesis at the end of 1991, the collection had moved to CD-ROMs and an online library. It was a special feeling to hold those images, knowing what they represented and what it took to collect them. I would spread them out on my table or the floor and Xerox copy them to make mosaics because I did not want to damage the precious prints.

Planetary exploration was beginning then, and Viking had a phenomenal impact on it. Its results shaped NASA's Mars exploration program vision up to the present. It gave us the first in-depth view of the history of a planet where everything looked incredibly familiar: ancient channels and dry lake beds, polar caps, dune fields, volcanoes, and lava flows now frozen in time. There is no need to invent new words to describe Mars. Its landscapes are very Earth-like and yet so different, a red planet with blue sunsets, where rovers have sunken their wheels in the dirt for the past couple of decades now.

In addition to the overall characterization of Mars, Viking 1 and 2

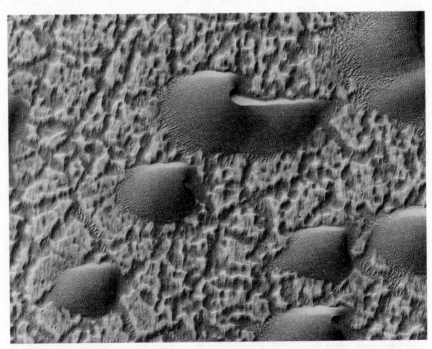

Barchan dunes in the southern hemisphere of Mars. These crescent-shaped dunes can reach between ten and three hundred meters wide and one and thirty meters in height.

landed with an extraordinary mission to accomplish. Led by Harold P. Klein, the biology team had four experiments on board each lander.[1] Their goal was to understand if life had developed and could still be present. The landers had to collect soil samples with their robotic arms and bring them to sealed containers for analysis to carry out the experiments. The Gas Chromatograph Mass Spectrometer (GCMS) investigation characterized the components of untreated soil released as the sample was heated at different temperatures. That experiment did not find any significant amount of organic molecules. Instead, the Martian soils seemed to contain less carbon than the lunar samples returned to Earth by the Apollo program a few years earlier.

The Gas Exchange (GEX) experiment investigated the hypothesis that metabolizing microorganisms could be living in the soil. Testing this idea required incubating soil samples at both sites. The Martian atmosphere was replaced with helium, and complex organic and inorganic nutrients and water were applied. The results came back identical and negative at both landing sites.

Then the Labeled Release (LR) experiment inoculated soil samples with a radioactive solution containing nutrients. If any microbes present in the soil were to process them, the air above the samples would show changes in radioactive carbon dioxide. To everyone's surprise, the data showed the release of radioactive gases immediately. Samples were then heated at temperatures ranging from 10 to 160 degrees Celsius and had substantially less to no radioactive gas release. The lead scientist of this experiment, Gilbert Levin—and others—remained convinced over the years that this was evidence of life on Mars, even though the results contradicted those of the GCMS and GEX experiments.

The last step was the Pyrolytic Release (PR) experiment, which tested the hypothesis that microorganisms would have adapted to Mars's aridity. The team heated up samples of Martian soil and measured if any gases related to life were released. The samples were left to incubate for several days, then gases were extracted, and the remainder of the soil was baked at 650°C. The Viking 1 PR experiment showed that a small amount of organic matter had formed, while the sterilized control did not. This result left open the possibility of biological activity, but issues were subsequently raised, since organics continued to be released at 90°C. Considering the lack of detection of organics in the ground by the GCMS experiment, the Viking science team ultimately concluded that a non-biological explanation was more likely.

Today, most of the planetary science community thinks that the Viking data are best explained by non-biological sources, such as oxidative chemical reactions from continuous surface exposure to short UV radiation. Mars has an ozone layer, which was discovered by ESA's Mars Express mission, but it is three hundred times weaker than Earth's. It comprises three different layers that vary substantially in location and time with a limited capacity to shield the surface. Non-biological sources create it as the sunlight breaks down atmospheric carbon dioxide molecules and releases oxygen atoms that recombine to form ozone.

The possibility of oxidative chemical reactions causing the Viking results found additional support in May 2008, when NASA's Phoenix mission discovered perchlorates. It is now generally accepted that such reactions explain what Viking landers observed, the presence of chlorinated hydrocarbons, which can break down organic compounds. Despite these observations, some still suggest that the Viking biological experiments lacked the sensitivity to detect trace amounts of organic compounds. As the samples were heated, perchlorates would have destroyed organics rapidly and produced chloromethane and dichloromethane, consistent with the Viking lander data. Even so, other studies reproduced the LR experiment results without involving any biological processes. Today, the experiment is still deemed inconclusive.

MARS, ALWAYS IN SIGHT

The Viking LR release experiment confronted the scientific community with some of the most frustrating issues related to identifying biosignatures beyond Earth. It was the first time then, but it would not be the last. The Allan Hills 84001 meteorite and, more recently, phosphine on Venus have since been added to the list of ambiguous results. Then something became apparent: in addition to the instrument sensitivity, the lack of background knowledge on the Martian environment in the mid-seventies prevented scientists from being able to put the results of the Viking biological experiments into context. This realization led NASA to a conclusion: if we were to search for life on Mars, it was imperative to collect as much information as possible on the past and present environment of the planet. As a result, NASA's Mars Exploration Program was born and continues to this day.

"Dingo Pass" at the Gale crater, Mars. It is, in my view, one of the most evocative images of our exploration of the red planet. It was taken on February 9, 2014, by the Curiosity rover. For scale, the dune is about a meter high and the camera is facing east.

And the American space agency is not the only one with its eyes set on Mars. In the sixties, the then Soviet Union became the first to attempt sending missions to the red planet, but never had much success with it. The Soviet missions account for close to 80 percent of all missions lost to Mars, but NASA, ESA, and others have not been immune to disaster. Over time, this secured Mars's reputation as a graveyard for spacecrafts.

The ongoing exploration of Mars is complex. It mixes the decades-long integrated visions of exploration of leading international space agencies with the first robotic steps of several nations at the surface of the red planet. It also includes collaborations between space agencies for future sample collection missions. There is no fixed date to bring these samples back to Earth, and the mission concepts are still evolving. NASA and ESA are working together on a plan to bring back rocks and dust for detailed studies between 2031 and 2033.[2]

The first phase of this mission is the Perseverance rover that landed in the Jezero crater on February 18, 2021. The rover currently collects samples and stores them in individual tubes, which are hermetically sealed and placed in the rover's belly. Perseverance left ten of them in a depot at a site named the Three Forks, a flat area clear of obstacles in the Jezero delta. There, they should be recovered in a few years by a sample return mission.

China is also planning a Mars mission, which would return samples to Earth by 2031. After developing a concept of a heavy-lift spacecraft carrying out all mission phases at once, China is now redesigning its concept to a multiphase mission, including a sample collection lander with an ascent vehicle. Like the NASA/ESA concept, an Earth Return Orbiter will send the samples back to Earth. Adding to this list of missions, Japan is making plans to retrieve samples from one of the moons of Mars, either Phobos or Deimos.

Meanwhile, goals focusing on the preparation for human exploration were added. Considering the challenges of landing and establishing a human outpost on Mars, understanding the planet's characteristics and dynamics makes perfect sense from the perspective of astronaut health, logistical needs, and in situ resources. Furthermore, if life is still present on Mars, the data can reveal the risks of forward and back contaminations and how to prevent them. This question is the focus of NASA's Planetary Protection program.

From the perspective of astrobiology, characterizing Mars's habitability from the moment of its formation to the present day is critical. It allows us to understand if simple life could have found favorable conditions to develop and survive. Since the nineties, missions in orbit and on the ground have helped generate a substantial body of knowledge. The mapping at a global scale by NASA's Mars Global Surveyor and Mars Odyssey orbiters focused on the past and present geology and climate history. They supported the landing site selection process for NASA's Spirit and Opportunity rovers that investigated Mars's past habitability by "following the water." The two rovers became the first to explore habitability from the surface, starting in 2004 at the Gusev crater and Meridiani Planum, where evidence of water was left in the ancient landscape's morphology and mineralogy. ESA's Mars Express joined the effort from orbit the same year. Unfortunately, Beagle 2, the landed element of ESA's mission, was lost. Two of the four solar panels did not deploy properly, blocking the communication antenna despite a successful landing.

International missions of increasing resolution and scientific capabilities followed, including the Phoenix lander, Mars Odyssey, Mars Reconnaissance Orbiter, Mars Science Lab, the Curiosity rover at the Gale crater, MAVEN, InSight for the U.S., ExoMars 2016–Trace Gas Orbiter for Europe, and the Mars Orbiter Mission (MOM) for India. These missions have since revealed the finer details of the evolution of the Martian geology, mineralogy, composition, atmosphere, climate, and tectonic activity. They were joined in Utopia Planitia by the Chinese rover Zhurong, which has stopped operating since then. NASA's Perseverance rover landed on February 18, 2021, and is currently exploring the Jezero crater. Mars 2020 is the first rover mission to search for past and present biosignatures in this ancient crater lake. Perseverance should be followed by ESA's ExoMars rover Rosalind Franklin, scheduled to land in Oxia Planum in the northern hemisphere, a landing site displaying one of the most extensive exposures of rocks and clays on Mars. Meanwhile, the launch of the European mission has been delayed by the war in Ukraine.

Owing to this vast volume of data, we now have a more accurate vision of how Mars evolved from an Earth-like planet to today's desolate world. The discovery of organic molecules hints that life could have originated early in environments and from processes similar to those of Earth. The detection of variable methane emissions is intriguing. While none of these observations represent unambiguous evidence of past or present biological activity by themselves, they do not contradict it, either.

Importantly, we now have robust evidence from all these missions that early Mars was habitable for (simple) life as we know it, but one of

Layered sedimentary rocks in the delta of the Jezero crater, Mars.

the most enduring uncertainties is the role that the unique early Martian environment may have played on habitability, organic molecules formation, and potential biosignatures preservation. Understanding how a biogeological record could have been transformed through the loss of the atmosphere, increasing biologically damaging ultraviolet radiation, cosmic rays, and chaotically driven climate changes is, therefore, key to unlocking where and what to search for on Mars, and also how.

A HABITABLE PLANET EARLY IN ITS HISTORY

If life ever started on early Mars, the planet would have provided natural shelters, liquid water, energy, sources of carbon, and nutrients. In addition, a much thicker CO_2 atmosphere than today's would have protected against UV. Although its current pressure is 165 times less than Earth's, studies of the upper atmosphere, its isotopic composition, and the analysis of Martian meteorites show long-term changes. Early Mars's atmospheric pressure may have been high enough to allow liquid water to flow and pond and to warm the surface temperatures. In contrast, the present residual atmosphere and range of temperatures only transiently reach a point where water's three phases (liquid, solid, gas) can coexist in thermodynamic equilibrium. The air pressure is so low that even in the most favorable locations, water can only be liquid between 0 and plus 10 degrees Celsius before it starts boiling. Nevertheless, periods of transient liquid water are still possible, especially with a high salt content, which depresses its freezing point, and may explain the seasonal reactivation of gullies and brine deposits.

Other possible shelters for life would have included the water column of lakes and episodic oceans, the subsurface and the interior of rocks, and the planet's deep interior. But for life to develop, energy sources would have been required. On early Mars, they could have come from sunlight (photosynthesis), chemical reactions (chemosynthesis), volcanic activity, and molecular decomposition by ionizing radiation (radiolysis). These are similar to energy sources found on Earth. NASA's Phoenix mission also discovered soluble minerals in the Martian dirt such as potassium, magnesium, and chlorides that could have provided nutrients for living organisms.

Mars seems thus to check all the boxes for habitability and life's ingredients but, when thinking about the possibility of life on the red planet, we cannot only stop at similarities with the Earth. We must also consider its uniqueness and how differences may have affected the formation of habitable environments, their duration, the development of life, and the preservation of its record. And here, despite the considerable volume of data accumulated so far, what we understand of Mars's geological history compared to the Earth can still be considered high-level "headlines," except at the landing sites, where rovers explore at a finer resolution. Even with the Opportunity rover, which covered the longest distance to date with 45.15 kilometers, the environmental variations at Meridiani Planum tell us mainly about a local story, granted that this story unfolded in the background of planetary-scale climate change.

The global perspective is gained from orbit, where spacecrafts help us decode the history of Mars, but the comparatively lower resolution of orbital data gives us only access to the major climate and geologic transitions. These transitions, taken individually, represent distant snapshots in space and time often separated by anywhere from tens of thousands to millions of years. In other words, we see a caricature of Mars's history, its general patterns, and outstanding features. It is almost like having completely different planets superimposed on top of each other. In reality, the tipping points we see are all connected through a continuum of environmental changes that cannot be seen from space, a continuum that explains how eons, eras, periods, epochs, and ages transition into each other. And, from the standpoint of life, what happens in that continuum is critical.

REVELATIONS IN A METEORITE

Despite these limitations, Mars is probably the world we know best next to ours. What we have learned so far helps us understand how its evolution may have shaped the fate of a hypothetical biosphere. In 2011, a 320-gram Martian meteorite was discovered in Rabt Sbayta in the Western Sahara desert in Morocco and nicknamed "Black Beauty." The meteorite's origin was confirmed by analyzing gas pockets within its minerals and their comparison with the unique signature of Mars's atmosphere, the composition of which was established by the Viking mission. This meteorite provided unprecedented insights into ancient processes that shaped the surface of the red planet and delivered a significant surprise. The analysis of decaying uranium trapped in zircon as the rock formed allowed scientists to determine the crust's age accurately. Its chemistry revealed the highest water content ever measured in a Martian meteorite. This water might have been derived from oceans that existed on early Mars and were still present when the rock was formed in volcanic terrain. The study of its composition demonstrated that the formation of Mars's core and the solidification of its magma ocean were completed extremely fast, within 20 million years of the formation of the solar system and no later than 4.547 billion years ago.

This discovery is significant from the perspective of a possible origin of life. It suggests that liquid water may have existed on the surface of Mars about 100 million years before it did on Earth. Considering the age of the first indirect evidence of life on Earth and the comparable environments between the two young planets, this discovery opens the possibility that life could have developed on Mars first. That does not prove it did, but the early Martian environment may have been amenable for life to start and take hold before our planet.

Importantly, these results provide an absolute age to anchor the chronology of Mars's environment. It is a critical step because until samples are returned from Mars, we can only obtain absolute ages from Martian meteorites like the Black Beauty. Otherwise, we rely on the count of impact crater densities on planetary surfaces to infer relative ages. These relative ages are then adjusted and calibrated to the lunar standard, for which we have absolute ages thanks to the samples returned by the Apollo program and, more recently, by the Chinese Chang'e 5 mission. Crater counts identified three main geological periods on Mars. A fourth one, the pre-Noachian, predates all of them, but has left little record. It is comparable in age to the Earth's Hadean eon, spanning between the accretion and the planet's differentiation from 4.1 to 4.5 billion years ago. Unfortunately, most of its geological record has been erased by erosion and high-impact cratering rates. However, one of the most characteristic features of the red planet was formed then and still dominates the landscape today.

Mars bears a strong asymmetry between its northern and southern hemispheres, which clearly separates the planet into two distinct topographic regions, with the old cratered highlands to the south and the young lowlands to the north. Various hypotheses attempt to explain its formation around 4.3 billion years ago. One is a collision with an asteroid nearly two-thirds the size of our moon (the mega-impact hypothesis). Another proposes that the accumulation of volcanic material in the lowlands would have caused the crust to sink under its own weight, forming the lowland region. Mantle convection was also suggested, a process that could have influenced the distribution of surface features and the elevation difference between the two hemispheres.

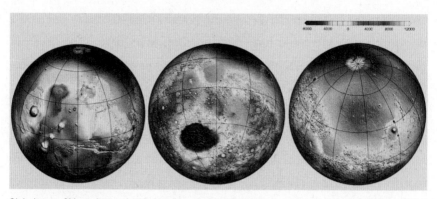

Global view of Mars obtained with data from the Mars Orbiter Laser Altimeter aboard the Mars Global Surveyor. The difference in altitude between the southern hemisphere plateau and the northern plains is clearly visible. The northern plains could have experienced episodic oceans during the history of Mars.

This asymmetry formed during accretion or right after, making it the most ancient geological feature of Mars. It is also one of the most critical features of the red planet. Its formation shaped geography and topography and influenced climate, hydrology, and geology, and it has to be considered central to any concept of a Martian biosphere.

The end of that period was marked by the loss of the magnetosphere, an event still recorded in the rocks. Then, around 4.1 billion years ago, the giant Hellas impact basin was formed in the southern hemisphere. Contrary to older formations, its rocks show a lack of magnetization, proof that the magnetic field that existed between 4.3 and 4.2 billion years ago had already shut down, leaving only remnants of it in the planet's weakly magnetized crust. Once the magnetic field was gone, the Martian atmosphere became rapidly eroded from sputtering by solar wind and radiation. From then on, it took only 300 to 500 million years for Mars to transform from an Earth-like planet to the arid desert we know today.

That period was followed by the Noachian period, which extended between 3.7 and 4.1 billion years ago. It was a time of peak hydrologic and geologic activity for Mars. Then, the red planet would have been the most similar to the Earth, providing comparable habitable environments if life had started. While river-like channels existed earlier, the Noachian period has the most robust record of them. Bodies of water and how they flowed were defined by the planet's asymmetry and the heavy bombardment of asteroids and comets that reshaped the topography billions of years ago. Impact cratering created watersheds and basins, where water and sediments accumulated. It also shaped the underground structure that formed the reservoirs, where aquifers stored water. The Noachian was the era of large lakes and inner seas, their record still visible in the southern cratered highland. The morphology of many river systems suggests that favorable conditions for their development were brief, likely due to strong climate oscillations induced by rapid changes in Mars's axial tilt. Although the atmosphere was vanishing due to the loss of the magnetosphere, these climate swings still triggered episodic periods of a warmer and thicker atmosphere through changes in Mars's orbit. At the same time, sustained volcanic eruptions released greenhouse gases and accumulated abundant sulfur deposits. Olympus Mons, the largest volcano in the solar system, and the other giant shield volcanoes of Tharsis began forming during the Noachian period. Countless other volcanoes were also active and displayed as many styles of volcanism as the Earth. Warmer periods are further supported by an abundance of clays produced through the interaction between water from streams, lakes, and hot springs with volcanic rocks, and through impact cratering.

DRY AND COLD, BUT CERTAINLY NOT DEAD

The transition between the Noachian and the Hesperian eons marked a tipping point, a time when Mars initiated its transformation into an arid and cold world. The Hesperian period spanned between 3.0 and 3.7 billion years ago. The large volcanoes of Tharsis and Elysium continued to build up and rise, and extensive lava plains were formed. In the same time frame, the landscape was deeply eroded by hundreds-of-kilometers-long catastrophic flows carved at the boundary of the highlands and the lowlands. They were sourced in vast areas of chaotic terrains that generated catastrophic water release, ending in the northern plains and forming episodic oceans. While residual shorelines are debated, the study of Hesperian deltas and possibly tsunami deposits support the idea of large volumes of water ponding for extended periods in the northern lowlands. As a result, episodic oceans or large lakes would continue to form sporadically during the next geological period, the Amazonian.

The Amazonian period has lasted for the past billion years, a reflection of how little Mars has been changed, but less activity does not imply that nothing has happened. Actually, quite a bit is still happening, albeit in small ways. Intermittent fluvial activity may have taken place in the last billion years, and possibly even more recently in the Elysium region, in an already hyperarid and cold environment. Although Mars has been geologically more quiescent, recent volcanic activity and volcanic-ice interaction prove that energy and water were still present only fifty thousand years ago in the Cerberus region in the northern hemisphere. The latest eruptions of Olympus Mons may have happened between 4 and 25 million years ago, possibly less, a blink of an eye on a geological timescale. As a result, the giant volcano is considered dormant, not extinct.

Mars still experiences quakes today. Since landing in November 2018 in Elysium Planitia, the Mars InSight lander has recorded over five hundred of them, most less than a magnitude 2 on the Richter scale. Many can probably be attributed to small asteroids and meteors crashing on the surface or simply explained by the cooling of the Martian crust. But others are larger (magnitude 3.1 to 3.6). Their origin was traced back to the Cerberus region as the epicenter of current seismic activity,

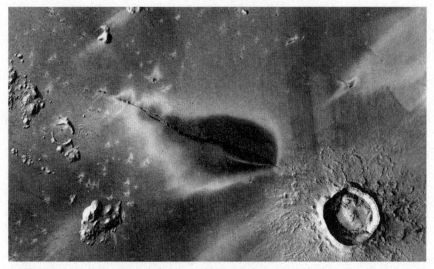

Recent volcanic deposits (fifty thousand years old) around a fissure in the Cerberus Fossae area on Mars.

consistent with the young volcanic deposits observed in that region from orbit. Occasionally, the spacecraft also captured larger quakes (4.1 and 4.2) originating from Valles Marineris. The largest recorded so far in that region was on May 4, 2022, and reached magnitude 4.7. Additional evidence from orbit led researchers in December 2022 to conclude that there is still a huge mantle plume of hot molten rock below the surface of Mars pushing upward in the Elysium region. This could explain both recent earthquakes and volcanism. So, no, Mars is not dead.

Transient flows of water still form today. They could originate from seepage of groundwater brines, outbursts of carbon dioxide, snowmelt, dry flows of windblown dust, or geothermal activity. The Martian surface reveals extended and varied remnants of past rock glaciers and debris-covered glaciers. Evidence of some of the most recent glaciations is still visible around the Olympus Mons volcano and might be barely 4 million years old, at most, and possibly as young as 350,000 years. When the planet's axial tilt passes thirty degrees, snow may also fall at the boundary of the lowlands and the highlands. A short-lived snowfall was actually observed by the Phoenix lander four kilometers away from the spacecraft landing site in the northern polar region in September 2008. But the most active agent of transformation of the landscape today remains the wind. It continues to reshape the landscape through the movement of ripples and dune fields, and regularly obscures the Martian sky by generating dust devils and dust storms.

This chronology of the evolution of the Martian environment is based on the geological record. Another timescale based on dominant mineral alteration caused by different types of weathering has also been developed. It was produced from ESA's Mars Express OMEGA instrument data, a visible and infrared mineralogical mapping spectrometer. This timescale is divided into three eras: the Phyllocian, which spans between the formation of Mars and 4 billion years ago, is an era named after phyllosilicates, or clay minerals, that characterized that time period; the Theiikian, which spanned between 3.5 and 4 billion years ago and was dominated by sulfate minerals deposited during intense volcanic activity; and the Siderikan (3.5 billion years ago to the present), named after the predominance of iron oxides. Both timescales have substantial overlap and equally highlight the dramatic changes that Mars underwent over its environmental continuum and its tipping points.

THE CARBON TRAIL

Along its own continuum, the Earth shows us that life interacts with its environment through changes to maintain a supply of energy and nutrients. In a biosphere, energy flows and is dissipated as heat, but chemical elements are recycled, and elements or compounds move between living and non-living forms and locations through biogeochemical cycles.

While this happens, life can significantly affect the processes controlling the exchanges between the atmosphere, the oceans, and the upper crust, leaving evidence of its activity.

Among those cycles, carbon and sulfur play an essential role here on Earth. And, if one of the primary goals of the Mars exploration is to "follow the water" to characterize habitability, it could be redefined as "follow the carbon" when it comes to the search for life. The reason this is so important is because carbon compounds are essential for organic matter and atmospheric gases. They also act as buffers in solutions, and they are critical ingredients in magmatic and volcanic processes. Sulfur, like carbon, has a crucial role and interacts across a network of reservoirs between the atmosphere, the surface, the ocean, and the subsurface through physical, chemical, and biological processes.

Considering the similarities between early Earth and early Mars, some of the processes that drove the geochemical cycles of carbon and sulfur on Earth likely operated on the red planet as well, at least earlier on. However, the rapid cooling of Mars negatively impacted the development of plate tectonics, which may never have started. If it did, it remained at a primitive stage, which could be an issue for habitability and life, since it plays a crucial role in driving the long-term carbon cycle. On Earth, it is responsible for carbon dioxide degassing at ridges and arcs and its return to the mantle through subduction. It also generates new, weatherable rocks on the surface. Instead, Mars was dominated by impact cratering, volcanic activity, and wind, which might have limited habitability. Despite these differences, atmospheric carbon could still have been recycled into the crust through underwater circulation and precipitated as carbonates, possibly explaining the lack of large surface deposits.

During magmatic activity, reactions between carbon and oxygen in the carbonate rocks and the hydrogen of the water produce methane (CH_4), which is then released into the atmosphere. Unlike the terrestrial carbon cycle, this process would have been less efficient as the atmosphere thinned and liquid water circulation and volcanic activity declined. Yet, methane is still produced on Mars today. It was detected from space and from the ground in the Gale crater with seasonal variations and occasional larger spikes. On Earth, 40 percent of methane is produced by living organisms. Since it breaks down in the presence of UV radiation, its lifetime on Mars should be three hundred to six hundred years, implying that the methane observed today cannot date back from 4 billion years ago, when conditions were more favorable. It has to be recent and renewed somehow, but it is still unclear how and by what process. Among the possible biotic sources, methane could originate from methanogenic organisms analogous to those found deep underground on Earth. They may have died off a long time ago but their decay could still generate methane, which could have been originally trapped in the frozen ground and then steadily released as the temperature warms up seasonally. Or microorganisms could have found a way to survive environmental changes on Mars, metabolizing methane in the subsurface, and they are the source of the current emissions.

Abiotic sources include iron oxidation in ancient volcanic and hydrothermal environments. Like in the former scenario, residual methane could have been trapped in the frozen ground and released later as environmental conditions changed. Methane may also form through a geochemical process called serpentinization. This process involves the interaction between heat, the mineral olivine, and water during hydrothermal activity, or through the interaction between UV radiation and comet dust falling on Mars. There is no shortage of options to explain methane on Mars, and at this time, all of them are still possible. As if it was not intriguing enough, in addition to methane emissions of unknown origin(s), Mars has given us another carbon mystery to ponder over recently.

Early in 2022, data from twenty-four sediment samples from the

Curiosity rover at the Gale crater revealed unusual carbon signatures. These samples included mudstones of an ancient lake. Using the rover's scientific payload, the remote science team on Earth collected them and baked them at 850 degrees Celsius to separate elements in the soil. Then pyrolysis was performed in a flow of inert helium to prevent any contamination. Methane was released during pyrolysis and analyzed. Among the samples, six showed elevated carbon 12 to carbon 13 ($^{12}C/^{13}C$) ratios. Data analysis showed seventy parts per thousand, which is a surprisingly large amount of carbon 12 compared to what is usually found in the Martian atmosphere or Martian meteorites. On Earth, living organisms use the lighter carbon ^{12}C to metabolize food or photosynthesize rather than the heavier ^{13}C. An abundance of ^{12}C is evidence of life-related chemistry on Earth.

But once again, as enticing as this result may be, several processes could explain it, and life is only one of them. Non-biological scenarios include a large cloud of galactic dust traversed by the solar system every hundred million years, which could have left carbon-rich deposits on Mars. Another considers the interaction of UV light and carbon dioxide in the Martian atmosphere that would have left molecules with the distinctive carbon signature on the surface. On the other hand, a biological origin could involve bacteria that convert carbon dioxide into methane. They are known on our planet to release methane in the atmosphere, which interacts with UV light to produce more complex molecules that settle back onto the surface. They are preserved in terrestrial rocks along with their carbon signatures. Could it be that Curiosity stumbled upon Martian life-forms akin to bacteria on Earth that convert carbon dioxide into methane? The only way to know is to continue documenting each of those hypotheses by collecting more samples, performing more tests, and seeing where the data take us.

As exciting as these new developments are, life is not only about carbon. Biology could have been sustained on early Mars using an element that was always abundant on the red planet. Many bacteria found in the sediments of terrestrial analogs to the red planet are microbes using sulfur as their primary energy source. Given the composition of the ground and the atmosphere from the Noachian into the Hesperian period, metabolism based on sulfur oxidation appears as a strong and viable candidate for producing energy for life on early Mars.

DEEP INTO THE MARTIAN CRUST

It took us a few decades of exploration and data gathering since Viking to come to this point. Today, we understand enough of the Martian environment, its plausible geochemical cycles, and how life could have interacted with them to infer the types of geological and chemical clues that these interactions could have left behind. These clues are the biosignatures we now search for with the most recent missions, including NASA's Curiosity rover and Mars 2020's Perseverance rover in the Jezero crater. Despite hostile conditions on the surface today, all data converge to show that Mars is on the high-priority list of worlds where life could have developed and survived over time. The new findings encourage us to think that we are on the right path.

Taking the optimistic stance and assuming for a moment that life appeared on early Mars, what would be the odds that it persisted to the present, and if it did, where would it be?[3] Obviously, it would be in environments that provide essential conditions for its survival. This could include underground habitats near or in reservoirs of liquid water. Such habitats would offer protection against the current hostile surface conditions and against cosmic radiations. In the absence of liquid water, ice could be a refuge for life as well. Ice is abundant on Mars, and not only at the polar caps. Vast reservoirs have been discovered in mid- to high latitudes recently.

Where they interact with salts and minerals, they could possibly create brine solutions supporting life. One of the holy grails in the search for life on Mars would be to find regions where water is still heated by magma, and where hot springs could still be present. These environments are central to all theories of the origins of life on Earth and the discovery of vast amounts of ice and recent volcanic activity leads scientists to believe that they may still exist. Research in terrestrial extreme environments also shows us that, provided that conditions for life are present, microbial organisms can survive in habitats as humble as rock pores and crevices.

Importantly, the notion of environmental continuum is central to our understanding of where life was and still could be today, not only for Mars but for any object in the universe. It comes into focus in the regions explored by the rovers. There, we have discovered that the evolution of Mars is not a black-and-white story such as the one we might narrate from orbit, but a story that has about as many shades of gray as our own planet, and this has critical implications for the evolution of habitability and life. The Perseverance rover just provided a few surprises that are good reminders of these nuances.

We landed in the Jezero crater to have a chance to study the evolution of a crater lake fed by a river that formed a distinctive delta. From

Residual delta buttes and volcanic blocks in the Jezero crater at a site informally called "Santa Cruz."

above, the story looked simple. Nevertheless, since landing, Perseverance has been revealing a surprisingly dynamic and complex hydrogeologic history in which environmental conditions may have reset many times. Perseverance has certainly found the evidence of an ancient river and a lake. It also found short episodes of sudden change, triggered by what may have been catastrophic floods that fed the forty-five-kilometer-wide crater to depths up to one hundred meters. Large boulders look out of place in the delta landscape. A few billion years old, they were formed from molten materials that cooled either below the surface or at the surface following volcanic eruptions.

Whatever their origin, these flows were capable of eroding the landscape extremely efficiently and rapidly. This detail matters because about a quarter of the Martian valley networks and channels seem to have been carved out from sudden events like these, rather than by prolonged-standing, slow flow discharges, changing environmental conditions for life. On the other hand, sudden flows and rapid and massive sediment deposits would be the perfect environment to entomb any life caught in these events. These discoveries make the exploration of Jezero extremely exciting, especially as the rover reaches the delta in search of biosignatures.

DID LIFE ON EARTH COME FROM MARS?

Why is there such resolve, almost an obsession with exploring what seems, for those who look too superficially, like a barren, cold, and hostile planet? The answer to that question may not be found in humanity's millennia of dreams and fantasies about Mars. It may be something much more profound, perhaps something personal and visceral, that may relate to the origin of life on Earth itself.

BLUE SUNSETS

When thinking about life on Mars, there are only two possibilities: it either emerged, or it didn't. Proving that it did not exist will be the most difficult. We need to think hard about what critical exploration steps we must complete before concluding that we must stop searching. We have just embarked on that journey, so the time is not now. But there is a deadline looming over the horizon that will have profound repercussions on our search regardless of whether life was ever on Mars or not. This deadline is set by the arrival of the first humans on the red planet, whenever this may be. Once we set foot on Mars, it will be only a matter of time before the planet becomes contaminated. Humans are walking microbiomes, and the environmental conditions on Mars might not be different enough to prevent some terrestrial microorganisms from adapting to a new planet. Humanity's arrival will thus signal a permanent modification of the Martian environment, and we may bring life to a place that possibly never had it before. In an alternate scenario, life could be different or somewhat similar to what simple life is on Earth, owing to the analogies between their early environments and the building blocks available to prebiotic chemistry and biology. But there might be more than just shared environmental conditions between early Earth and Mars. We may discover that life on both planets presents many similarities because there is a reasonable probability that they could share the same roots. It is all a matter of orbital mechanics and timing.

When they formed, both Earth and Mars were constantly bombarded by large comets and asteroids, many of them kilometer-sized planetoids. Violent collisions ejected debris out of the gravitational field of both planets. Then rocks traveled up to millions of years in space before finding their way to each other's surfaces. There is no doubt that Mars and Earth exchanged material that way. We have clear evidence that Martian rocks made it to the Earth with the Martian meteorites.

While it theoretically works both ways, Mars has a few things

89

THE SECRET LIFE OF THE UNIVERSE

conspiring to make it easier for Martian rocks to journey to Earth. Being smaller and much less massive than the Earth, Mars's escape velocity is only 45 percent that of the Earth's. Its atmosphere is also 165 times thinner than the Earth's—granted that it would have been thicker earlier on. In other words, it takes a lot less energy for debris to escape Mars than it takes for it to escape the Earth. Remember also that Mars's crust cooled about 100 million years sooner than the Earth's. And the Black Beauty meteorite tells us that water was already present on the red planet by then. Assuming prebiotic chemistry transitioned to biology on Mars as fast as it did on Earth, life could have emerged 100 million years earlier on the red planet. There is a chance, thus, that some of these Martian rocks could have carried complex organic molecules or even primitive organisms to the Earth. A life that originated on Mars may have found a favorable environment to evolve and thrive here on Earth from that point on. Ultimately, we might be the Martians we so relentlessly strive to discover with our spacecrafts and rovers. And, as we prepare to travel to the red planet, we might, in fact, just be making the journey back home.

Whether Mars was the original cradle for life on Earth or not, there is another reason why the red planet is so unique and why its exploration is so important to us. The record of our origin is gone from the surface of the Earth—forever lost to geological recycling. Most of the Hadean and early Archean rocks have been erased by plate tectonics and erosion. This forgotten time is Earth's "biological event horizon" beyond which we cannot access our origins. Comparatively, terrains of the same ages have been better preserved on Mars. Even if Mars has a distinct origin of life, the similarities between its early environment and our planet could reveal some fundamental clues about how prebiotic chemistry transitioned to life here on Earth. Mars holds the Rosetta stone of our origin in more than one way, and the answer to how we came to be might be waiting for us in an ancient Martian outcrop.

5

PLANETARY SHORES

The Pale Blue Dot: the simple idea of it evokes visions of Earth, most of its surface covered by water pulsing blue in the darkness of space. Over the millennia, oceans and seas have been synonymous with humanity's wanderings, of journeys of exploration to uncharted lands as we leave the safety of a harbor to set sail into the unknown. They speak of the vastness of our world and this endless curiosity that inspires humans to always reach beyond the horizon. It is thus no surprise that Carl Sagan used them as a metaphor to evoke what was then the infancy of space exploration: "We began as wanderers, and we are wanderers still. We have lingered long enough on the shores of the cosmic ocean. We are ready at last to set sail for the stars." As he wrote these lines in Cosmos, his metaphor was coming to life in our own planetary neighborhood.

OCEAN WORLDS

In 1979, Voyager 1 and 2 gave us the first clues that the surface of Jupiter's moon Europa was young and active when they revealed a highly

On September 29, 2022, the Juno spacecraft gave us the first view of Europa since the Galileo probe twenty-two years earlier. This image of its icy, fractured shell was captured about 1,500 kilometers from the surface.

reflective icy crust covered with just a few impact craters. Years later, the Galileo mission returned images of Europa's jigsaw puzzle–like surface, reinforcing the suspicion that a subsurface ocean was lurking down below. Salts and water vapor in the moon's atmosphere hinted at chemical exchanges between this hidden ocean and the surface.

Meanwhile, the Cassini mission discovered and mapped over a hundred plumes shooting up 200 kilometers into space on Enceladus, a small moon of Saturn. Racing over 65,000 kilometers per hour, the spacecraft passed just 48 kilometers over Enceladus's south pole, where ice particles mixed with water vapor and organic materials were jettisoned 1,200 kilometers per hour into space from a region named "the tiger stripes." There parallel faults in the moon's crust displayed an anomalous heat signature that provided yet another sign of liquid water at depth, warmed by hydrothermal processes.

This discovery was just the beginning. We soon realized that the Earth is only one of many ocean worlds in our solar system, most of them hidden from view under thick covers of rock and ice, except for Titan, where large exotic seas and lakes first suspected in the Voyager 1 and 2 data were later confirmed by Cassini, which observed them at the surface. Some of these ocean worlds have been established since, and many more are suspected. All of them hold tremendous potential for life despite their remote location from the sun.

FROZEN KINGDOMS

Deep into our planetary neighborhood, beyond Mars and the asteroid belt, begins the cold realm of the outer solar system. From Jupiter, the sun is 5.2 times farther away and smaller than it is from the Earth. With a few exceptions, most moons orbiting the gas giants look frozen in time. Yet, appearances can be deceiving. Over the last decades, the exploration of planets and Earth's extreme environments have transformed our views on habitability. Worlds far too distant from the sun to be considered in the habitable zone may still experience environmental conditions beneath their surface that could lead to the emergence of life. Reinforcing their astrobiological significance, deep oceanic hydrothermal vents on Earth harbor complex ecosystems that survive without being bound to the surface for energy, metabolism, or nutrients.

The oceans on these moons may or may not be global. Some are better candidates for life than others, but whether any alien biospheres arose on these far distant worlds still hinges on the universal mantra of water, energy, nutrients, shelter, and carbon sources. It also depends on the sustainability of biogeological cycles over time. Except for dwarf planets Ceres and Pluto, most of them orbit in the lethal radiation environment of gas giants. Astronauts in their space suits exposed to the surface of Europa might survive fifteen minutes. They would last only a few seconds without them in the vacuum of space, subjected to enough radiation to fry anything living to a crisp. Radiation from Jupiter can destroy molecules at the surface of its moons instantly, and, with perhaps the exception of Titan, the largest moon of Saturn, life would not survive on the surface of any of these worlds. The same applies to biosignatures and chemical evidence that would be destroyed if left unprotected.

If life exists on these worlds, it will be found within their icy crusts or deep interiors. These environments are their ecosystem boundaries. Since the sun cannot be an efficient energy source, these potential biospheres may have to draw energy primarily from the interior. Relentless gravitational tides constantly destabilize the interior of these moons, deforming them in daily ebbs and flows that grind rocks and fracture their icy shells. The resulting tidal friction generates enough heat to power hydrothermal activity within the oceans. The radioactivity of the rocks and the interaction of galactic cosmic rays with the icy shells of the moons can produce energy through radiolysis. Nutrients created by this process could rain down on the oceans deep below through fractures and faults. This connectivity between the surface and the deep interior is an essential element of a resilient biosphere and one that could make the difference between the emergence of life or a lifeless ocean world. And life's presence fosters dispersal, adaptation, and the diversification of habitats, all of which increase its chances of surviving and accessing new sources of energy and nutrients.

Such hypothetical biospheres would be shaped by environmental variability over the short and the long term, which would also drive the location, structure, and nature of life's habitats and evolution. Tidal forces are examples of variability and cyclicity induced by gravity. Since these forces are omnipresent, they should play a primary role in the distribution, state, and evolution of life's habitats. They must also play a role in the loops and feedback mechanisms of ecosystems where gravitational tides, habitat selection, and the metabolic activity of microorganisms are intertwined.

But tides are not the only players. Part of the process also involves the characteristics of these oceans. Unlike on Earth, most of them are permanently locked within the interior of the icy moons, and changes in their characteristics cannot occur through interactions with the atmosphere. Instead, they happen at the interfaces between water and ice and between water and rock on the seafloor, which results in water density, temperature, and salinity changes. On Earth, water circulation and evaporation are driven by wind, sunlight, and currents. Motion and mixing are essentially driven by convection from the seafloor through thermal circulation on the ocean worlds where solar energy is lacking. Ocean motion is also affected by competing forces between the moons' inertia and their rotation, which generates Hadley-like circulation, where warmer water rises toward the midlatitudes and cooler water sinks at the poles. Importantly, circulation and mixing are essential for the distribution of nutrients in an ecosystem. Identifying what they are and what triggers them is key to finding where alien life may be located and characterizing ecological rhythms, these biological and environmental processes that repeatedly occur at definable intervals.

These oceans lay hidden from view, so their presence can only be inferred from indirect evidence. For example, variations in the thermal signatures of the surface show the existence of hot spots that may help keep water liquid in the interior. Plumes and geysers show that liquid water moves around within the icy crust, and tidal stress keeps them spewing upward. Young faults and fracture systems, the lack of impact craters, and the filling of basins with fresh surfaces are other indicators of tectonic activity and newly deposited materials. The topography and the high reflectivity of an icy crust provide more evidence that a young and active surface is undergoing temporal changes. Salt deposits in the proximity of fault systems add further mineralogical clues to the existence of salt water encased in an icy shell. The presence of a thin atmosphere can also sometimes, but not always, indicate communication between the surface and a subsurface ocean and the escape of gas through fractures, faults, and chaotic terrain. But not all clues are so readily visible, and some demand more complex measurements, such as tidal effects, rotational disturbances, and the detection of gravitational anomalies.

MANY CANDIDATES

Many oceans may exist in the outer solar system, but their presence is more probable in some worlds than others. Such is the case for Enceladus, Europa, Titan, Ganymede, and Callisto. Among these, Europa and Enceladus show evidence of this communication between the surface and the ocean so critical to habitability. The thick icy shells of Ganymede and Callisto may preclude it. Titan is a different world altogether and must be treated in another class of its own, and we'll discuss it in detail in chapter 6. Beyond these five moons, Neptune's satellite Triton, Saturn's moon Dione, and dwarf planets Ceres and Pluto are considered candidate ocean worlds based on limited spacecraft observations.

The Uranian moons are shrouded in even more mystery. While Voyager 2 flew by Uranus, it did not come close enough to the moons to detect any telltale signs of subsurface oceans. Yet, images from the spacecraft clearly showed the presence of active volcanism, but not of molten rock like on Earth. These worlds are so cold that the warmest material erupting from their volcanoes is water ice! They also show signs of recent geologic activity. Of Uranus's twenty-seven moons, five of them, Ariel, Miranda, Oberon, Titania, and Umbriel, may be ocean worlds.

Even confirmed ocean worlds are not born equal, and there are many characteristics that affect their chances of developing life. For instance, an ocean might be hiding deep beneath the rocky and icy surface of the Jovian moon Callisto. One of the four moons first discovered by Galileo in the early seventeenth century, Callisto is tidally locked to Jupiter, orbiting around it in the same amount of time it takes Jupiter to rotate. As a result, it always presents the same face to the giant planet. It is about two and a half times smaller than the Earth and the most heavily cratered world in the solar system. The abundance of craters shows that very little geological activity has occurred since its formation 4.5 billion years ago. What we see today are mostly unchanged surfaces formed billions of years ago. There is no sign of active volcanoes or tectonics eroding the craters. Only bright white spots cap their central peaks with water ice. Callisto has an extremely tenuous atmosphere composed primarily of carbon dioxide with some oxygen in its exosphere at the edge of space. An atmosphere this thin should be lost in about four days if it is not renewed regularly by the sublimation of carbon dioxide ice from its icy crust.

Callisto doesn't have a magnetic field that comes from within the moon itself, but it behaves as though it has a layer that conducts electricity, which could be consistent with a salty ocean. If this ocean exists, it is locked away 250 kilometers beneath the surface without any visible communication with it. An ecosystem developing in this ocean would have to rely solely on a closed system where rock, ice, water, and heat interact to produce all essential elements of biology. This system may become depleted sooner than an ocean where communication with the surface exists. As of today, Callisto remains something of a mystery. Nevertheless, it could represent an extreme case in the spectrum of ocean worlds, and its exploration may teach us critical lessons on the limits of habitability and life in those environments.

Ganymede raises some of the same questions but presents possibly a better case for life. The largest—5,266 kilometers in diameter—and most massive moon of the solar system is a fascinating world with a metallic core, an internal ocean, and an icy crust. After Galileo, Juno was the latest spacecraft to return data from the giant moon on June 7, 2021, as it flew only one thousand kilometers away from its surface.

The best evidence of a subsurface ocean did not come from these planetary probes. They came from the Hubble Space Telescope in 2015. Ganymede is the only moon known to have a magnetic field, which causes aurorae to form. Because the moon orbits so close to the gas giant, it is also caught in Jupiter's magnetic field, whose fluctuations affect the

Close-up view of Ganymede captured by the Juno spacecraft during its flyby on June 7, 2021.

development of the aurorae, and no effect would be detected if the moon did not have an ocean. Instead, astronomers using Hubble noticed a rocking motion of the aurorae, evidence that something was fighting back against Jupiter's magnetic field. Like Callisto, this "something" is likely a large amount of salt water beneath Ganymede's crust that explains the disturbance. This ocean could be located 150 kilometers under the icy crust. It is a staggering ninety kilometers deep, thus possibly ten times deeper than Earth's oceans, and contains more water than all terrestrial oceans combined.

The morphology of this moon's surface reinforces the conclusions from Hubble. Ganymede has large and bright regions of ridges and grooves that cut across older, darker terrain. Their formation is explained by global tectonic processes that put the active surface under tension. The older terrain is flooded in places by ice volcanism that covers old surfaces with smooth-looking young material. The moon is surrounded by a thin atmosphere of oxygen produced when space radiation breaks down water ice on the surface into hydrogen and oxygen. Hydrogen escapes faster into space due to its low atomic mass, leaving oxygen behind.

Thus Ganymede has plenty of water, a magnetic field, and a source of energy generated by tidal heating that triggers tectonic activity. But is this enough for the development of life? It is too early to tell. There are still too many open-ended questions to say whether life had a chance on Ganymede. First, we need a better understanding of its ocean characteristics, structure, topography, and the nature of the layers that confine it in the subsurface and at depth. Additional data may help us determine the possible communication pathways between the ocean and the surface. At this point, there is no clear evidence that such pathways exist. If this is confirmed, it could limit the influx of nutrients to potential organisms locked in an ocean below a very thick ice shell, just like with Callisto. As with all systems, it is a question of balance between the sources and the sinks. To some extent, we can approximate what takes place deep beneath the surface of Ganymede through modeling, and we do so by using our knowledge of Earth's processes and the physical and chemical characteristics of ocean worlds we gather with our spacecrafts and telescopes.

For Ganymede, where communication between the atmosphere and the ocean might not exist, nutrients would have to be generated within the oceanic system itself and from hydrothermal vents on the ocean floor. Other sources of nutrients may come from minerals extracted from the weathering of rocks. Elements essential to life, like phosphorus, would be depleted over short timescales in alkaline oceans with hydrothermal activity. These conditions do not preclude the existence of life, but they suggest it could exist only in low concentrations and might not be easily detectable. Obviously, these conclusions are for life as we know it in environments as we know them.

But, as always in exploration, that may not prevent Ganymede from surprising us. ESA's Juice (Jupiter Icy Moons Explorer) mission launched in April 2023. It will spend at least three years exploring Jupiter and its system to focus on the study of Ganymede as a potential habitat for life. Juice will also provide additional information about Callisto and Europa as part of a comparison between the Galilean moons as more evidence is needed to build a solid case for their habitability. Meanwhile, other icy worlds continue to lengthen the list of potential candidates, including, most recently, Mimas, a moon of Saturn.

MYSTERIOUS ENCELADUS

Of all these candidates for life in the outer solar system, Europa, Enceladus, and Titan are the most promising. We all marveled at the discovery of Enceladus's geysers. The moon is now a fixture of our imaginations, and its dynamic landscape is familiar to many. It is thus hard to realize that two puzzling and long-standing questions were finally solved only recently for this small moon of Saturn. One relates to its reflectivity— Enceladus is the brightest object in the solar system. It reflects close to 100 percent of the light that hits it. The other involves its relationship to Saturn's E ring. The Cassini mission delivered the answer, which happened to be the same for both questions.

Before space exploration, astronomers already knew something was different with Saturn's E ring. Compared to the planet's other rings, it is not as sharply defined. Its extraordinary dimensions span between 120,000 and 420,000 kilometers above Saturn's equator. This vast, yet diffuse, disk is composed of ice and dust spread between the orbits of Mimas and Titan. Sitting squarely in the E ring's center, Enceladus orbits 386,000 kilometers away from Saturn. Its interaction with the E ring was suspected long before the beginning of planetary exploration. Another hint came from the brightness of the icy moon, which meant that Enceladus could not be made of dirty ice, the kind that has been exposed for too long to the vacuum of space and cosmic radiation. Very early on, astronomers also noted that its south pole was even brighter than the rest of the globe but could not figure out why. Cassini put an end to the questioning in 2005, when the spacecraft returned its first images of Enceladus, revealing over one hundred geysers. The particles they spew escape either into space, where they contribute to the formation of the E ring, or fall back on Enceladus's surface, continuously renewing it with a cover of fresh snowy material. Part of this material also makes its way to neighboring moons Mimas and Tethys, increasing their reflectivity. This observation was only the beginning of a series of discoveries that would force us to rethink where habitability and life might be possible.

Barely five hundred kilometers across, Enceladus is by all measures a diminutive moon and, for most, looks like a scratched, moon-sized snowball. Despite its small size, it is fully differentiated with a silica core and a water-rich mantle. Salts and silica dust are being forced upward by geysers into the overlying icy crust in the so-called tiger stripes region near the south pole. Their composition show that they form in at least $90^{\circ}C$ water interacting with rocks. By contrast, the surface of Enceladus is extremely cold, colder than any of the other Saturnian moons, with a temperature of $-198^{\circ}C$ at noon. It is covered in cratered regions and smooth plains bordered by ridged terrain only a few hundred million years old. Other areas were recently resurfaced and do not show any impact craters. Fissures, fractures, and cliffs testify to the intense tension and deformation the moon is experiencing.

All these observations finally found an explanation in early 2005 as the plumes of Enceladus were discovered in the Cassini images. The spacecraft subsequently flew through them to analyze their composition. The later flybys detected plumes extending up to five hundred kilometers into space, continuously erupting from pressurized chambers and resembling the activity of terrestrial geysers and fumaroles (vents that release steam and volcanic gases). More recently, JWST detected one of these plumes spraying nearly 10,000 kilometers away from the small moon. Overall, their intensity varies as a function of Enceladus's position in its orbit. It is greater when the moon is away from Saturn than when it is closer, a telltale sign that tidal stress plays a role in opening and closing fissures in the crust. But if the tidal effect from Saturn were the only contributing factor, Enceladus would become frozen entirely to the core within 30 million years.

The display of energy observed on the small moon requires an enormous heat source, over one hundred times more than what the natural decay of radioactive elements in rocks can produce. And it would not

Geysers on Enceladus, imaged by Cassini on November 30, 2010, from 67,000 kilometers away. South is up.

explain why the heat source is concentrated in the tiger stripes region. On the other hand, these observations fall into place if Enceladus has a porous core enabling water to permeate it easily.[1] This process would allow cold water from the ocean to seep into the core and gradually heat through tidal friction. As it becomes heated, the warmer water rises, transferring the heat to the base of the ocean, where water vents through narrow plumes, carrying small particles from the seafloor within them. Transfer from the seafloor to the surface takes weeks to months, and then they are released into space through the geysers.

A model based on Enceladus having a porous core was produced by Nantes University researchers in France in 2017 and satisfied all the observations, including the global ocean. Their computer simulations also showed that most water would be expelled from a narrow region around the south pole, with localized hot spots and a thinner ice shell just above, which is precisely what the data from Cassini show. With this porous core scenario, Enceladus would have enough heat to power hydrothermal activity through tidal friction for billions of years. Since sustainability is paramount to a biosphere, this strengthens the idea that this ocean world exhibits conditions suitable for life.

Enceladus revealed more hidden astrobiological treasures as Cassini passed through the geysers. From Cassini, we learned that the ocean is slightly salty, and hydrothermal activity is sustained on the ocean floor in contact with the rocks. Complex organic molecules were also discovered, likely processed in this hydrothermally active core. They could have been injected into the ocean by hydrothermal vents in a process reminiscent of scenarios for the emergence of life on Earth. How these complex organic molecules ended up propelled into space is the result of another process known on Earth. They simply hitchhiked a ride on gas bubbles coming from the vents on the ocean floor. They burst at the interface between the ocean and the icy crust as they rise, dispersing the organics in droplets that become ice-coated when vapor freezes on their surface. Then, as fissures open following tidal cycles, they are ejected in the plumes. Sulfur and nitrogen were also identified in forms that life can use, and the porous core model demonstrates that Enceladus has steady energy sources. More intriguing is the unexpectedly high level of methane detected in the plumes that known geochemical processes cannot solely explain. It is consistent with microbial activity near a hydrothermal vent, or non-biologic processes, different from those known on Earth.

EXPLORING OCEAN WORLDS

The habitability of Enceladus's ocean has been established, and concepts for future missions there are being designed. One of them is the

Enceladus Vent Explorer (EVE).[2] It is being tested in terrestrial glacier crevasses as an analog to the depths of Enceladus and possibly Europa's plume vents. EVE is a study phase concept that, if selected as a future mission, would be sent into the active geysers to reach the subsurface ocean. Exploration could be performed by one or a swarm of self-propelled eel-like robots that use geysers as a pathway toward the ocean through the ice shell.

Other studies are currently developing technology concepts that could be reused for all ocean worlds. The SWIM mission concept is one of those.[3] The acronym stands for Sensing with Independent Microswimmers. Like the Enceladus Vent Explorers, it is in the study phase and originates from the Jet Propulsion Laboratory. SWIM would deploy up to hundreds of untethered, individually controllable swimming microrobots into a planetary ocean or lake. Such a mission could find biomarkers and signs of habitability in ocean worlds, including Europa, Enceladus, and the lakes and seas of Titan.

The first in-depth investigation of a potential ocean world will come from the Europa Clipper mission[4] set to arrive at Jupiter in April 2030, where it will characterize the habitability of the smallest of Jupiter's Galilean satellites, Europa, which is 90 percent the size of Earth's moon. Soon after, it will be joined in the Jovian system by ESA's Juice mission. Europa Clipper will perform dozens of close flybys, gathering information about Europa's icy shell and ocean and possibly flying through plumes, as Cassini did with Enceladus. Unlike Enceladus, plumes are elusive on Europa, and they may also have multiple origins. The first hint of their existence came from the Hubble Space Telescope in 2012 and then again in 2014. The images appeared to show water vapor plumes shooting up two hundred kilometers into space. Repeated observations indicate sporadic eruptions in between calmer periods.

FASCINATING EUROPA

The recent discovery of Europa's plumes inspired researchers to re-examine the 1997 Galileo flyby data, in which they found evidence that went unnoticed decades ago, simply because analytical tools were not sophisticated enough at the time to make sense of them. The Galileo data revealed a localized bend in the magnetic field that had never been explained, up until a generation later: indeed, Galileo had caught a plume in action. Some large plumes may originate from the communication between the subsurface ocean and the surface, like Enceladus. Smaller ones could develop from pockets of water embedded higher in the icy shell. These pockets may form following the impact of asteroids and comets and result from the tremendous heat generated from these collisions.

The twenty-nine-kilometer-wide Manannán crater gives us an example of what can happen. The impact that formed the crater 20 million years ago generated heat that melted the ice in the surrounding region, creating a water pocket that expanded and moved sideways. As it did, it accumulated salt from the crust. With time, the pocket progressively froze. The remaining water became pressurized, causing it to erupt. It left behind "spider-like" marks as evidence of ancient geyser activity not too dissimilar from those observed in the south pole of Mars when CO_2 erupts in the spring. While these deposits hint at shallow bodies of liquid water within the icy crust itself, the ocean lies deep beneath it. The first clues of the existence of an ocean on Europa were obtained from the Voyager and Galileo images.

The moon is tidally locked to Jupiter. Complex gravitational disturbances from the gas giant and the other Galilean satellites maintain a slight eccentricity in its orbit. It causes Europa to oscillate and experience a stronger pull when it comes closer to Jupiter than when it moves away from it. This oscillation deforms the moon, and the resulting tension is released through the formation of streaks, cracks, and faults. A small number of impact craters signifies that the surface is active and relatively young. It is also covered in chaotic terrain, dark bands whose opposite sides match each other, and large blocks resembling pieces of a jigsaw puzzle. These cracks have dark, icy material that recently flowed into opened gaps.

Europa's geysers may come from deep ocean water and closer surface pockets. Deposits of salts at the surface are bombarded by sulfur originating from the moon Io's volcanism, explaining the observed spectral signatures.

Most patterns in the terrain can be explained as a response to stress generated by the tidal changes as the moon orbits Jupiter. But some of them do not fit this model, particularly the longest, one thousand kilometers long, features for which a process somewhat similar to plate tectonics on Earth is at play. It is as if the surface of Europa moved independently from the rest of the moon's interior, which could be possible if a layer of liquid, or slightly warmer ice on which the icy shell rests, is squeezed between the crust and the deep interior. Moreover, Jupiter's magnetic field shows disruption in the space around Europa, suggesting an induced magnetic field created within the moon by a deep layer of electrically conductive fluid.

Like other moons, the best candidate is a global ocean of salty water that could be 160 kilometers deep and somewhere between 10 and 32 kilometers beneath the icy surface. Europa Clipper has a ready experiment to test this hypothesis. If an ocean lies underneath the crust, the pull and push of Jupiter's gravity should deform the surface by about thirty meters. If the moon is completely frozen to the core, the deformation should only be about one meter.

Europa is, beyond a shadow of a doubt, one of the most fascinating and promising worlds in our search for possible abodes of life in our solar system. Although we still need a lot more information to grasp its full potential, we can try to sketch out what a plausible environment for life may look like with the information already available. At the surface, the temperature never rises above $-160°C$ at the equator and $-220°C$ at the poles. The base of the ice shell should be at the freezing point of a hypothetical ocean lying below. Depending on pressure and salinity, that temperature probably ranges between $0°$ and $-4°C$ where the ice and the ocean meet. The heat sources for water and life on the ocean may come from radioactive decay in the moon's interior and tidal dissipation from the interactions between Europa and Jupiter and the other Galilean moons. These heat sources combined are likely to generate intense convective currents. Convective plumes from the relatively hot (about $4°C$) oceanic interior coming into contact with the icy shell would cool down and descend into the ocean. This process may also erode the base of the icy shell over time. Several geological features, particularly the chaotic terrains made of regions of fractured ice, could be areas where deep, warm ocean waters have come into contact with the overlying ice and where the icy shell briefly melted through.

In another scenario, the surface of Europa, like Venus, could be periodically experiencing a catastrophic renewal of its surface. This overturn would take place every 5 million years. Perhaps not coincidentally, this is within the range of the estimated age for the surface of Europa. The potential thus exists for the ocean to not only communicate with the surface in areas of Europa where the icy shell is thinner and fractured, but also to mix with it regularly.

The promise of remarkable discoveries in these various environments is tantalizing. Shallow pockets of water may provide critical clues about the geochemistry of the icy shell and the volume of organics stored in it after being delivered by comets and asteroids over time. Europa already had all the essential ingredients for life as it formed (carbon, hydrogen, nitrogen, oxygen, phosphorus, and sulfur). With comet and asteroid collisions, the chemical elements for life could be found both within the crust and in the ocean and move from one to the other through aqueous environments within faults, cracks, and chaotic terrain. They could also be mixed during periods of crustal overturn. Tidal heating may be powering a system that cycles water and nutrients between the crust and the ocean almost indefinitely, maintaining a nutrient-rich, energy-rich environment conducive to life and its sustainability. And then what? What if we find life in Europa's ocean? Many would tell you that they would be shocked if we did not discover life on Europa. In truth, this small moon, just like Enceladus, has everything we can hope for in our quest for life beyond Earth. If they are correct, then what type of life could it be?

Conservatively, microorganisms are observed in somewhat analog environments around hydrothermal vents at the bottom of terrestrial oceans. So microbial life appears to be a baseline scenario for life on the icy moon, if it exists. But recent studies go beyond the idea of microbes and propose that Europa's ocean could contain enough oxygen to sustain fishlike animals! Further, the observation of sources of carbon by JWST has important implications for the habitability of Europa's ocean. Carbon is a biologically essential element and the backbone of life as we know it. Regardless of what it is, if it exists, life on Europa is bound to bring a wealth of new knowledge, particularly since it will have little chance to be related to life on Earth. Unlike Mars, planetary exchanges with these icy worlds are improbable. Here again, celestial mechanics dictate the rules. Even with a steady flow of material coming from the Earth or Mars, the transit time would be about 2 billion years, which microbes might not survive. Even if they do, and depending on trajectory, the impact velocity would be between 18,000 and 112,000 kilometers per hour upon arrival. The lower range might be survivable, but not the higher end.

These observations tell us that if life emerged on Europa and Enceladus, it would be an indigenous life, a second genesis. It would not be biology seeded by the Earth or Mars. We could learn so much from such a fantastic discovery, particularly about the many ways life can assemble itself in the universe from its simplest building blocks. Yet, it could still be life as we know—or at least recognize—it, but this could change with Saturn's largest and most intriguing moon: Titan.

6

TITAN: A WORLD
OF UNKNOWNS

A small group of us had gathered in the hall near the main audi-
torium of our building at NASA Ames in California. We were
anxiously waiting. This was in the late morning on January 14, 2005,
and ESA's Huygens probe was about to land on Titan. If it had gone according to script, the lander would have been already safely sitting on
the surface of Saturn's largest moon, Titan, by the time we met in the
hall. Up to that point, the last view of Huygens we had received was an
image captured a little over three weeks earlier by Cassini shortly after it
had separated from the probe. Now Huygens was reaching its final destination. Hopefully, it had successfully made it through the atmosphere
suspended from its parachute, and we would all soon discover Titan up
close and personal for the first time. The last opportunity to explore this
region of the solar system had been with Voyager 2 in August 1981, and
the moon was captured by the spacecraft's cameras from over 4.5 million
kilometers away.

Titan had been known to astronomers for far longer, however. It
was discovered in March 1655 by Christiaan Huygens. The presence of
an atmosphere was first suspected in 1903 by Spanish astronomer Josep

Comas i Solà and later confirmed by Gerard P. Kuiper in 1944. Then, Pioneer 11 and Voyager 1 gave us the first closer look at the moon, sending intriguing data back to Earth. On its way out of the solar system, Pioneer 11 passed nearly 584,000 kilometers away from Titan on September 1, 1979. The camera resolution could not define more than a distant and fuzzy orange world from that distance. A couple of months later, on November 12, 1979, Voyager 1 came within 4,000 kilometers of the moon, revealing a surface hidden by a thick, nitrogen-rich atmosphere. It was thus no surprise that when plans for the next mission to the Saturn system were conceived, the exploration of Titan was front and center. After joining forces, NASA, ESA, and the Italian Space Agency (ASI) were about to deliver what remains one of the most extraordinary and most successful flagship missions in the history of planetary exploration. NASA provided the Cassini space probe, ESA the Huygens lander, and ASI contributed Cassini's high-gain antenna, the majority of the radio system, and parts of the science instrumentation. Cassini was destined for a long journey of exploration in Saturn's system that ended on September 15, 2017.

FIRST IMAGES OF TITAN

On that day in January 2005, all eyes were locked on the TV screen. Huygens was set to descend through Titan's atmosphere, acquire critical data on its composition and structure, and land safely. And so we were waiting. The excitement was steadily building in our little group moments before the first image appeared on the screen. Some of my colleagues grabbed a few chairs and sat next to a large table. I was standing; my neck was already aching from looking up for too long at the black television screen bolted up to the wall. We knew there had been communication issues between Huygens and Cassini, the latter serving as a relay with the Earth. Some of them were caught early, and work-around solutions were implemented, but with only one communication channel working, about half of the images were lost.

Once the download started, the images came in bursts. They had been captured on the descent. Huygens snapped the first one we saw about sixteen kilometers from the surface as it finally came out of the haze. I was dumbfounded then, like I am today, and every time I look at these images again. We could see hills, riverbeds, possibly a shoreline, and a darker surface. To my eyes, it looked almost exactly like the landscape I fly over when I come back home from my travels abroad as the plane is about to land in the Bay Area near San Francisco. The only difference was the apparent lack of vegetation or the hustle of humanity and its omnipresent footprint on the landscape. These images offered a glimpse into the reality of a wondrous world that would give astrobiologists a chance to flex their knowledge and imagination.

Over eighteen years have passed since ESA pulled off this spectacular landing on Titan. Huygens was the first spacecraft to touch down on this uncharted moon and the first to land on a world in the outer solar system. It was an extraordinary comeback for ESA after the loss of Beagle 2 one year earlier, a resounding triumph that would allow humanity to discover new and uncharted planetary shores.

Most of the Huygens mission took place during its 153-minute descent into Titan's thick atmosphere, the thickest of any moon in the solar system. Once on the ground, the probe lasted another seventy-two minutes before it ran out of power and contact was lost. But critical scientific discoveries were made in that short amount of time. After the Pioneer and Voyager flybys, many questions were raised about Titan's atmosphere, particularly the origin of the nitrogen and methane. The high levels of methane represented a mystery; sunlight should have destroyed it. Thanks to Huygens, we can now shed some light on a number of these questions. In the process, we made a few unexpected discoveries that have profound implications for the potential of Saturn's largest moon to harbor life.

A PECULIAR ATMOSPHERE

During its descent, the probe collected data on the structure of the atmosphere, its temperature, pressure, and density. Above 500 kilometers, Huygens measured temperatures of $-100°C$, fluctuating locally due to inversion layers, gravity waves, and tides. Extremes $(-87°C$ and $-203°C)$ were observed at the upper and lower boundaries of the stratosphere. Meanwhile, the surface registered at $-180°C$ with an atmospheric pressure nearly 1.5 that of Earth. Titan was also veiled in a thick opaque haze blocking most visible sunlight. Huygens's descent was far from smooth. The Doppler Wind Experiment and the analysis of the descent imagery showed the probe drifting east by 160 kilometers, pushed by 400 kilometers per hour sustained winds.

Later observations by Cassini suggested that Titan is a "super-rotator," its atmosphere rotating much faster than its surface in the same way Venus's atmosphere does. Farther down, the lander was shaken by strong wind shears before finally touching down and sliding over a short distance in a much calmer environment. Despite the rough ride, Huygens returned the first direct data on the high-atmosphere composition, isotopic ratios, and other trace gases, including organic compounds. It confirmed that Titan's atmosphere is primarily made of 97 percent nitrogen and 2.7 percent methane with trace amounts of other gases. The composition of the atmosphere, including its isotopes, indicated that nitrogen might come in part from materials in the Oort cloud and not from the Saturn system. The Oort cloud is a sphere of sorts composed of frozen and inactive comets located between 2,000 and 200,000 astronomical units (AU) from the sun. Comets may thus have been incorporated early into Titan's core, including dense, organic-rich material.

In the upper part of the atmosphere, methane levels were low, but reached saturation around 6.5 kilometers altitude. Then Huygens recorded a sudden 40 percent increase. At the same time, nitrogen remained constant, suggesting the presence of liquid methane locally on the ground, possibly due to the probe heating the surface. While methane may have multiple origins, carbon isotopes did not indicate that microbes produced it. They were more likely the result of trapped ice beneath the surface during the moon's formation. This scenario could explain why methane is still present. Under current conditions, it should be destroyed in the atmosphere within a few tens of millions of years through photochemical processes if it was not replenished somehow. Stored as ice in the planet's interior, methane might be released episodically into the atmosphere through cryovolcanism.

Titan's atmosphere is also rich in organic molecules. Hydrocarbons are produced by a chain of chemical reactions above nine-hundred-kilometer altitude in the atmosphere. Sunlight and high-energetic particles break down nitrogen and methane molecules at this altitude, allowing complex organic chemistry. Some molecules coalesce into larger ones that sink into the lower atmosphere. The higher density facilitates the formation of even bigger aggregates that produce the haze of carbon-based aerosols surrounding the moon. In theory, complex organic compounds like tholins may form on Titan, and polycyclic aromatic hydrocarbons (PAHs) have been detected from Earth-based telescopic observations. They played a significant role in the origin of life on Earth by mediating RNA synthesis. Since then, closed-loop molecules, including C_3H_2

(cyclopropenylidene), were also detected. Discovering tholins was significant, as they form the backbone rings for the nucleobases of DNA and RNA on Earth. It is the simplest and smallest closed-loop molecule found in any atmosphere, suggesting that weird prebiotic chemistry could occur. On Titan, these organic molecules are raining down on the surface!

Huygens also carried out an experiment to search for evidence of lightning. While none was observed, extremely low-frequency signals were detected, which led to the discovery of a conductive salty ocean of water and ammonia fifty to eighty kilometers below the nonconductive icy crust. By then, we already knew that weird prebiotic chemistry could be taking place and possibly give rise to life as we do not know it on the surface, but the existence of liquid water at depth opened new possibilities. Titan increasingly looked like the perfect alchemist's cauldron. It represented a natural lab, a world similar to early Earth, where we could study the chemical processes that had led to life over 4 billion years ago on our planet. It could also allow us to start cutting our teeth on a completely alien, methane-based biochemistry. But that was not Huygens's role. Its mission focused on the atmosphere, and the data it returned contributed to revealing the full potential of Titan as a possible abode of life.

RAINING METHANE

As the probe descended into the atmosphere, the radar initiated data acquisition 1,200 kilometers above the surface. It revealed a bright highland sitting 100 meters higher than the darker dissected terrain where Huygens landed, which resembled terrestrial lake beds and alluvial surfaces. The Descent Imager/Spectral Radiometer offered a chance to finally discover for the first time what lay hidden beneath the orange haze. In those images, drainage networks appeared to cut into one-hundred-meter-steep ravines very similar to terrestrial riverbeds, draining from the plateau into the darker lowland region. Their shape suggested rapid erosion from sudden and violent flows. Others seemed formed by rainfall or sapping by liquid methane. These networks likely contributed to forming a lake not too far from Huygens's landing area, but the basin did not show any sign of ponding liquid methane at the time of the landing.

Unlike the Mars rovers, the Huygens probe could not rely on high-resolution data for the selection of a landing site before the mission or much data of any kind to begin with. The probe was thus designed to survive both a hard landing or a splash landing in one of Titan's seas and last long enough to collect some data on the ground. After drifting in the wind, Huygens ended its journey, settling down near the Adiri region on a flat surface of damp sand or packed granular ice particles. The

Left: Titan and Tethys seen by the Cassini spacecraft on November 26, 2009 (source: NASA/JPL/ Space Science Institute). Right: Huygens discovers rounded pebbles and a grainy frozen surface on January 14, 2005.

landscape looked eerie yet familiar, extremely alien and Earth-like, with rounded cobbles in the foreground. Everything else was flat.

The lander took the same image repeatedly until it died, capturing in the process the formation of a dewdrop of methane caused by the heat from Huygens and likely explaining the spike in methane concentration recorded by the instruments. I was convinced from the first sight of the rounded rocks that Huygens had landed on a beach. Later analysis suggested that the probe probably came to rest on a riverbed covered with cobbles made of hydrocarbons and water ice. Rocks made of solid water ice rolled and rounded by time in torrents of liquid methane. To me, that alone epitomizes the weirdness of Titan.

In the short time of its mission, Huygens produced a fantastic wealth of data,[1] and gave us a better understanding of the complex physical and chemical processes in the atmosphere. It also pointed to new and fascinating research directions in a world where everything looks familiar, yet nothing is really what it seems, and where we could be given a chance to explore side-by-side life as we know it and life as we don't.

After the end of Huygens's mission, it was time for Cassini to pursue the investigation from orbit. After a journey of seven years from the Earth to Saturn and the separation from Huygens, Cassini initiated a thirteen-year-long scientific odyssey in orbit around the Saturn system, a journey of remarkable discoveries where Titan remained a strong focus. During that time, Cassini completed over one hundred close flybys of the giant moon, mapped its surface, and continued to make detailed studies of its atmosphere. The detection of large gravity tides seemed to confirm the presence of a layer of liquid water underneath its icy crust, kilometers below the lakes and seas of methane sitting on the surface. In the deep interior, water is kept from freezing by internal heat produced by the interactions of Titan with Saturn and the other moons. Methane-based chemistry could also act as an antifreeze. Data suggest that the ocean could be as salty as the Dead Sea and likely made of a brine of water mixed with sulfur, sodium, and potassium. Despite this, the variable thickness of the icy crust may be a sign that the interior ocean is in the process of freezing. If true, this could impact its habitability and how methane is being returned from the crust to the atmosphere. Ice volcanism may occur only in localized hot spots and not through a global process similar to plate tectonics on Earth if the crust is mostly stiff.

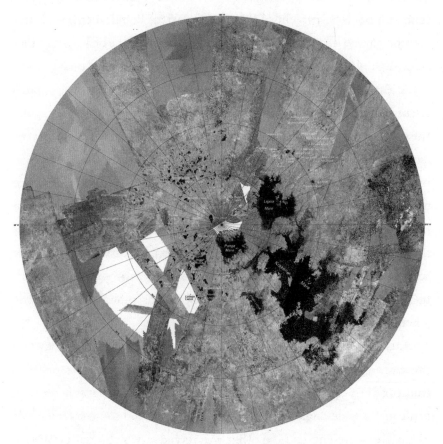

A map of Titan's northern hemisphere, generated from Cassini radar data (2004–2013), displays methane and ethane seas and lakes. The north pole is at the center, and the view goes up to 50N Latitude. Dark areas represent the seas and lakes, with small lakes (<30 km) visible above and to the left of the north pole. Nearly 97 percent of these features are located in the northern hemisphere, in an area covering 900 by 1,800 kilometers.

Like the hydrological cycle on Earth, Titan has precipitation and liquid collecting in the crust in aquifers and streams and rivers that flow downhill. It ponds in basins to form lakes and seas, most of them located in the polar regions. Oases may also exist in the more arid equatorial region and form after occasional rainstorms. The largest sea on Titan is located in the northern polar region. Named Kraken Mare, it is almost as large as the Caspian Sea and mostly composed of liquid methane and parts of liquid ethane and nitrogen. The sea has drawn particular attention, as it holds 80 percent of the moon's surface liquid. It covers almost 520,000 square kilometers, and radar data reveal depths of 85 to 305 meters. High winds generate ripples and waves at the surface and, Titan being less massive than the Earth and with a weaker gravity, these waves are seven times taller and three times slower than on our planet, while the moon's orbital eccentricity leads to tides as high as one to five meters. Seas and lakes evaporate, ice volcanoes erupt, and, as a result, clouds form in the atmosphere. As condensation takes place and rain falls, the cycle can begin again, but it is driven by methane instead of water, unlike on Earth, and, similarly to our planet, the weather is very much driven by seasons. Titan's year is nearly as long as thirty Earth years, so its four seasons last 7.5 years each, while equinoxes and solstices occur every fifteen years. They are characterized by dust storms, monsoonal rains, and downpours that fill basins and shape alluvial fans.

Cassini unveiled the secrets of Titan's surface and the geology and processes hiding beneath the thick haze in two ways. One was with its Imaging Science Subsystem, which consisted of two cameras (a narrow-angle and a wide-angle) sensitive to visible light and some infrared and ultraviolet wavelengths. The other was its radar, which was the instrument that truly penetrated the thick and murky atmosphere. The combination of their datasets finally gave us a better idea of what Titan looks like from the ground.[2] Nearly two-thirds of its surface is covered in flat plains. Vast fields of individual dunes reaching up to a 1.5-kilometer-wide and 100-meter-high stretch across the equatorial region over hundreds of kilometers. Unlike their terrestrial counterparts, these dunes are not made of silica sand. Instead, they are composed of solid water ice coated with hydrocarbons, which form when radiation hits the ice and chemically reacts to create organic molecules directly at the surface. They could also form in the atmosphere and fall as rain or both; we do not know for sure.

Titan has hills, mountains, and terrain carved by rain and erosion dubbed "labyrinths." The rest of the moon is the domain of lakes and seas. Together, Huygens and Cassini revealed the astounding diversity of Titan. They helped us identify its different environments, including the thick, organic-rich atmosphere where a complex, alien organic chemistry is brewing. There is also the geologically active surface, where a liquid cycle mimics the hydrological cycle on Earth, with seas and lakes filled with liquid methane. Titan has a decoupled icy shell and global subsurface salty liquid water ocean. In it, ammonia and other salts produce buoyant liquids that bubble up through the crust and free methane from the ice. Beneath the ocean, a layer of high-pressure ice may be present, and underneath it, a watery silicate core. These are Titan's environmental domains. To understand whether a biosphere may exist or if it is even possible, we must now set them in motion and see how they interact with each other and what is produced through these interactions.

LIFE IN A FROZEN SOUP?

Titan's environment represents the closest analogy to early Earth's "primordial soup" we will ever find in the solar system, except that it is ten times farther away from the sun than our planet and receives one hundred times less energy. So as far as the surface is concerned, that soup is frozen and represents a forbidden land for life as we know it. Drop any terrestrial microbe in one of Titan's lakes and it will instantly look like Han Solo in the carbonite slab of *The Empire Strikes Back*. But not everybody is ready to give up on the possibility of life at the surface of Titan, especially in its lakes and seas, where it could use methane as a solvent instead of water. Methanogenic life could take in hydrogen in place of oxygen, react with acetylene instead of glucose, and produce methane instead of carbon dioxide. Ways to test this idea have included the measurement of hydrogen concentrations in the atmosphere and acetylene levels on the surface. Intriguingly, both measurements returned results consistent with the possibility of life, but mineralogical or meteorological processes could also explain them.

For any life to survive on Titan, it would have to be extremely strange by any terrestrial standards. For instance, it could not be held together by membranes made from fatty molecules called lipids, which would not survive in Titan's conditions. Protocells could forgo the lipid part and use another molecule, acrylonitrile (CH_2CHCN), whose unique properties let a molecule attract another and form a membrane. This molecule has been discovered in abundance in Titan's atmosphere, enough to support millions of single-celled life-forms theoretically. So, could acrylonitrile-based cells survive on Titan?

Computer simulations based on a pure methane environment demonstrated how floating acrylonitrile molecules would act while bumping into molecules of methane at minus 183 degrees Celsius. Results show that temperature might prevent the formation of flat and flexible sheets required to wrap up a cell on Earth. Instead, molecules would organize in rigid crystal-like structures comparable to salt or ice. Under those circumstances, living molecules on Titan might skip membranes altogether and use the frigid environment to hold together and stick to rocks. For instance, they would wait for the wind, currents, and tides to bring nutrients. To some extent, it would be a passive way of living supported by a dynamic environment. It would be a somewhat different type of coevolution of life and environment from what we experience here on Earth and one that might not bring as many options to evolution. The adaptation and survival tools of such life, its dispersal pathways, and ecological cycles are open-ended questions, but long seasons with varied weather and storms would likely be essential components, including the regular input of organic material from the atmosphere.

The possibility that weird life arose on Titan is a story that continues to unfold. It has gained momentum over the years with additional data from Cassini, Huygens, ground-based telescope observations, and modeling and computing capabilities. Today, researchers are warming up (so to speak) to the idea of cold life as we do not know it, at least in lakes and seas, in an otherwise extraordinarily challenging environment on the surface. But things may be different at depth.

SCENARIOS FOR LIFE IN A SUBSURFACE OCEAN

The discovery of the subsurface water ocean is a game changer.[3] While it is not an indicator of life in itself, it brings Titan's prebiotic chemistry and potential biochemistry back into the domain of liquid water, and at temperatures that are survivable for simple terrestrial life. To evaluate its potential as a habitable environment, we must figure out the possible pathways organic molecules that formed in the atmosphere and on the surface could take to reach the interior ocean. If these pathways exist, organic molecules could serve as building blocks and chemical nutrients for life at depth. Understanding how their exposure at the surface modifies them could also help us develop biosignature detection strategies.

It could also provide insights into how exposure at the surface modifies them and how they are transported from the surface to the ocean.

In this "top-down" scenario, organic compounds produced in the atmosphere fall to the surface via precipitation or direct airfall. Once at the surface, their degree of alteration depends on where they are deposited, their residence time, and the type of transportation agents involved, which can be wind, rivers, and precipitation. Ultimately, part of these organics finds their way into the methane lakes and seas, where they settle at the bottom of basins as a sludge layer. From there, they might be delivered to the interior ocean through transport in the fractured ice shell. Once in the ocean, they are distributed with inorganic material in the water column and deposited in bottom sediments.

But this process does not have to be necessarily completely a top-down journey from the atmosphere to the ocean. Here, previous findings on the origin of the atmosphere of Titan may come into play again. For example, comets originating from the Oort cloud that were integrated into Titan early in its formation are rich in organic materials. Since this material is already part of the moon's deep interior, it could enter the ocean from below. As a result, even though organic molecules cannot reach the ocean from the surface, the ocean would still contain the building blocks of life in a "bottom-up" pathway, from the moon's interior to its ocean. In the best-case scenario for life, both pathways would be possible.

Titan's subsurface ocean may be in contact with the ice-rock core, possibly allowing oxidation-reduction (redox) gradients and the production of heavier elements, both of which are critical for a habitable environment and the production of energy sources for life. On Earth, redox gradients in bottom sediments are created by variations in mineral composition and characteristics of the water column. While the nature of

Titan's seafloor is not precisely known, it could be made of an interface between high-pressure ice and rock. Although it is not a perfect analog, one of the most extensive ecosystems on Earth lives in somewhat similar conditions, without light or oxygen, just beneath the ocean floor. Called the dark biosphere, this ecosystem lives in the ocean crust. It is composed of a broad diversity of microorganic communities, and some of them could be relevant to Titan. They feed on inorganic molecules created during the chemical alteration of rocks by the water, and they release methane as waste. Others, like ammonia-oxidizing bacteria, can survive in the dark, oxygen-depleted ocean environments by producing oxygen on their own through the conversion of ammonia into nitrite, a process they use to metabolize energy. To achieve this, they need to produce their oxygen along with nitrogen gas as a by-product.

The most widespread ecosystem on Earth is found in the oceanic crust, and most microbial ecosystems on Earth exist in the dark. Through their activity, they change the biogeochemical cycles and the chemical composition of the oceans. An analog process might be taking place on Titan. The only way to know will be if the signatures of the chemical changes associated with their metabolic activity can make it back to the surface through cryovolcanism. While no life that originates in the deep ocean is expected to survive at the surface, ice volcanism could be depositing biosignatures from this deep life for surface explorers to search for, and that time has finally come. No other spacecraft has landed on Titan since Huygens, but this is about to change with NASA's Dragonfly mission. Dragonfly will be launched in June 2027. The cruise vehicle will deliver the probe to the surface in 2034 after a 1.29-billion-kilometer journey.[4] Its goals include searching for chemical biosignatures and investigating Titan's active methane cycle. Dragonfly will also explore the prebiotic chemistry taking place in the atmosphere and on the surface.

DRAGONFLY

Of all the worlds in the solar system, Titan is the one that offers the greatest range of possible exploration strategies. Robots can land or roam on its solid surface and make a splash landing with lake landers. They can also explore the subsurface and depths of vast seas and lakes with submarines, drift in the atmosphere with balloons, or fly with robotic airplanes. Flying is easy on Titan, with an atmosphere 1.5 times denser than the Earth. The Ingenuity helicopter on Mars has spectacularly shown the way in the much thinner Martian air by successfully logging numerous demonstrations and scouting flights for the Perseverance rover.

Dragonfly was preceded by several concepts that made it through the study stage but were ultimately not selected as missions. However, these concepts demonstrated the flexibility and wide range of technological approaches enabled by Titan's environment. Over the years, they included a variety of proposals for hot-air balloons, alone or with landers and orbiters. One such project was to fly an airplane powered by advanced radioisotopic generators. In 2010, the Titan Mare Explorer (TiME) project was designed to explore Ligeia Mare, the second-largest sea on Titan. The idea was to use a lake lander to take measurements of the composition of the sea and perform imaging, meteorological observations, and sonar surveys. This concept was one of three Discovery program finalists, but the Mars InSight mission was ultimately selected instead. I still believe today that TiME was one of the most creative concepts I have seen in many years, and I hope it will sail one day in the not-too-distant future at the surface of Ligeia Mare.

But, for now, the spotlight is on Dragonfly. This concept, led by Zibi Turtle, is in the same class of outstanding projects as TiME. Both projects merge scientific rigor, technology innovation, and the intangible magic and romance of some missions that make us feel that we are living through the pages of a science fiction novel.

Dragonfly will take advantage of the thick atmosphere to fly a drone-like probe and explore Titan for at least three years. This probe is not exactly your backyard drone, either. It is a 450-kilogram, drone-like quadcopter/octocopter, using eight rotors in four pairs to fly from one location to the next on Titan. It is about the size of the Perseverance rover, which is three meters long, two meters wide, and 2.7 meters tall. Dragonfly will be delivered from space in an aeroshell and will descend through the atmosphere with a parachute. Once the parachute is jettisoned, the drone will land under rotor power and deploy a high-gain antenna for direct-to-Earth communication. The probe has a radioisotope power supply that provides heat and energy. During the three years of the nominal mission, the large drone will sample the surface near the equator of Titan, and search for telltale signs of biological activity near

Selk impact crater (ninety kilometers in diameter) on Titan, located at 7N and 199W, the landing region planned for the Dragonfly mission.

THE SECRET LIFE OF THE UNIVERSE

the Selk crater, in the dune fields of the Shangri-La region. The proximity of Selk could allow us to study subsurface material exposed in the debris field left by the impact.

Dragonfly will use a mass spectrometer to identify chemical components relevant to biological processes in surface and atmospheric samples. A gamma-ray and neutron spectrometer will characterize the surface composition under the lander, and a geophysics and meteorology package will study the weather and carry a seismometer. The Dragonfly camera suite comprises microscopic and panoramic cameras to image the terrain and scout for compelling landing sites. Engineering and monitoring instruments may also be used to determine the characteristics of the atmosphere and interior. The radio link via Doppler and ranging measurements will assist in the analysis of Titan's rotation state, which is influenced by its internal structure.

Beyond Dragonfly, other mission concepts are currently under study, including an insulated submersible that would characterize organics. Carbon and nitrogen are abundant on Titan, but whether life functions can be performed in a solvent like methane is unknown. One advantage of such a mission would be to enable access to bottom sediments and possibly discover chemical, physical, and biological processes that do not occur at the surface. The targeted exploration site, like for TiME, is Ligeia Mare, a nearly 127,000-square-kilometer sea with 2,000 kilometers of shoreline. The sea possibly connects to Kraken Mare through a channel, which could be explored during an extended mission. The overall mission's range could be as much as 1,200 kilometers in twelve months, half of the time used for science. The submersible itself would come through Titan's atmosphere, protected in an aeroshell delivered by the orbiter that would also support operations. Methane provides a transparent medium for communication with the submersible, avoiding the need for resurfacing.

Dragonfly will investigate Titan's surface composition in different geological settings, explore habitability, assess prebiotic chemistry, and search for water- or hydrocarbon-based life.

While the Cassini-Huygens mission has allowed us to answer some questions about Titan, thirteen years of exploration have raised many more. Dragonfly is designed to help us test all the predictions we made about how Titan may work as a system, a potentially habitable world, and maybe a biosphere for both life as we know it and life as we do not know it. Titan is an extraordinarily fascinating world. But beyond the fascination, it gives us a chance to have a first taste of what alien life may look like, something that may be entirely foreign from everything we know. It is a natural lab within the confines of our solar system, a backyard to take our first steps toward understanding what life on faraway exoplanets may look like and, importantly, how to search for its signatures.

7

NEW
HORIZONS

With the ocean worlds, we discovered that liquid water is not unique to Earth, but common throughout the solar system, bolstering our hopes of finding living oases away from the habitable zone. Even so, the most enthusiastic scientist would limit how big or far away from the sun we are ready to believe a world could be and still meet the criteria for habitability. Despite it all, Enceladus, the small moon of Saturn, turned out to be one of the best prospects for finding life in the solar system. But, then, within four months of each other, the Dawn and New Horizons' missions pushed back theoretical limits even farther.

SIZABLE SURPRISES

Dwarf planet Ceres came into view out of the darkness of the rocky no-man's-land between Jupiter and Mars in March 2015. It was the second stop on the Dawn mission's journey through the asteroid belt after its exploration of Vesta. Dawn was to study these two small worlds to answer questions about the solar system's formation. About a quarter the size of our moon, Ceres (941-kilometer diameter) represents on its own 25 percent of the asteroid belt's total mass. It is a relic of sorts, a remnant of what protoplanets would have looked like a little over 4.5 billion years ago, and a witness of the solar system's deep time.

It is also a reminder of its unruly youth, when planets were moving about before they reached their current locations, ramming through space, pushing and pulling on each other in monumental tugs-of-war best explained today by the Grand Tack and the Nice theories. In those days, the solar system did not look anything like it does today. Jupiter was on its way toward the sun until it was pulled back to its current position by the formation of Saturn. This early event brought chaos, propelling many protoplanets toward the sun to meet a fiery demise and most likely ejecting a few others out of our planetary neighborhood altogether. It also prevented Mars from becoming as big as it should have been, possibly the size of the Earth and Venus, and changed its fate entirely.

As for Ceres, its composition tells us that it was not formed in the asteroid belt at all, but more likely near the orbit of Neptune and then migrated inward. This suspected connection with the icy worlds was echoed by the discovery of ammonia salts and sodium carbonates in the ninety-two-kilometer-diameter Occator crater. Ceres is still today under the gravitational influence of the gas giants. They cause cyclic shifts in its axial tilt, making it oscillate between two and twenty degrees, and trigger seasons episodically, the last time fourteen thousand years ago. During episodes of maximum axial tilt, impact craters left in the shadow act as cold traps, and they may retain water ice over billions of years, like the moon and Mercury.

Just as for its origin, defining what type of world Ceres is precisely has been a struggle. First considered a planet when it was discovered by the Italian astronomer Father Giuseppe Piazzi in 1801, it was later reclassified as an asteroid when more objects were found in the same region of the sky. Then it was caught in the dwarf planet drama and, because of its planet-like characteristics, was designated as a dwarf planet along with Pluto and Eris, but due to its location in the asteroid belt, it sometimes retains a dual appellation as a dwarf planet and an asteroid. This indecision reflects the sum of knowledge accumulated in the past decades of exploration, and the realization that our classification of planetary objects must be fine-tuned and broadened, in the same way stars are classified by types that reflect their specific characteristics and evolution. Despite its size, Ceres is a complex world showing signs of recent geological activity.

With enough heat available early in its history, its interior was partially differentiated into an icy mantle and a liquid water ocean. The existence of a small core is, however, still in question. If the ocean had frozen over time, a thick layer of ice should have formed, which Dawn did not find. Instead, a subsurface body of briny liquid might still be present. If it is not global, it lingers at least in pockets. An indirect clue about the relative proportion of ice left in the subsurface was given by the shape of a few large impact craters that suggested a resurfacing through slow collapse of their walls and ice volcanism. Another evidence is Ceres's density, consistent with 25 percent ice by mass. Its water content is only second to the Earth's. Everything else is rocks, hydrated salts, and clathrates, compounds where gas molecules are trapped in a cage of water molecules.

Variations in the gravitational field also point to a subsurface reservoir made of salty water, possibly hundreds of kilometers wide and forty kilometers deep. Water vapor produced by the sublimation of near-surface ice detected by ESA's Herschel Space Observatory is another indicator of an icy subsurface. It might also show the oceanic slush permeating to the surface through faults and fractures and sublimating into space.

The discovery of Ahuna Mons, a one-of-a-kind, 4.8-kilometer-high and 19-kilometer-wide mountain, added geological evidence. Its location is on the opposite side of Ceres from the planet's largest impact crater, the 283-kilometer-diameter Kerwan basin. The seismic energy of the impact at Kerwan radiated and focused on the opposite side of the dwarf planet, fractured its crust, and facilitated the upwelling of muddy water ice and sodium carbonate salts from the ocean layer. Ahuna Mons is only a few hundred million years old and will probably disappear in time, like other mountains on Ceres did before. Cryovolcanoes were formed throughout Ceres's history, but the viscosity of the material they are made of stretches them outward, flattening them over a billion years, ultimately leading to their destruction. Those observed today are only in, or near, young impact craters and formed relatively recently. Nevertheless, their presence shows that Ceres remains geologically active, with brines still present near the surface.

Then there is also the dome located at the center of the 22-million-year-old Occator crater named Cerealia Facula. The 96-kilometer-diameter impact basin has the brightest deposits discovered on Ceres.

The mosaic of Cerealia Facula (fifteen kilometers) in the Occator crater is composed of images taken by Dawn at a distance of thirty-five kilometers above Ceres's surface and projected onto a topographic model (without vertical exaggeration).

Initially observed by Hubble in 2003, these bright spots and several others perplexed scientists and remained mysterious until the Dawn mission finally revealed their origin. At Occator, where they are the most spectacular, they cover both the central region and other localized areas on the crater's floor. They almost looked like lights beaming into space when Dawn's camera captured them in its field of view for the first time. Some of them could be as recent as 2 million years old. Made of magnesium sulfate brines, aluminum, ammonia-rich clays, and sodium carbonates, their formation required hydrothermal activity and temperatures above 50 degrees Celsius. Occator is not the only object on Ceres where they are observed. The surface is dotted with hundreds of them, predominantly found on the peaks and pits of young impact craters, their floors, and their rims and walls. They formed due to the heat generated by impacts, the upwelling of volatile-rich material through the fractured crust, and their precipitation at the surface.

Ceres is the only dwarf planet in the inner solar system and, in many ways, a unique world. Its interior is likely in the process of freezing, but it is not frozen just yet, as shown by its recent geological activity. It also checks all the conditions to make it a potentially habitable world. It is about 50 percent water by volume, with a deep reservoir of salt-enriched water sheltered beneath a fractured icy crust. Some of the minerals observed on Ceres have also been detected in the geysers erupting from Enceladus, and their composition shows that sources of energy still exist. Cryovolcanism may be active today as well. Organic molecules, including tholins, either produced internally or delivered by comets, were discovered by Dawn in and around the fifty-one-kilometer Ernutet crater and other patchy deposits across the surface of Ceres. Their organics content could reach as much as 40 to 50 percent of the spectral signal observed in Dawn's data compared with meteorites found on Earth. Moreover, the detection of carbonates and clays raises the possibility that they were processed in the same warm and water-rich hydrothermal environment known to be favorable to the emergence of life. Clearly, Ceres is an intriguing small world. Its ocean and the presence of organics are not proof that life ever existed or does exist today, but its building blocks and the chemistry for its development are unexpectedly present.

PLUTO'S EXTRAORDINARY WORLD

As surprising as this was, 2015 would deliver another shocker. It took only four months after the beginning of the Dawn mission for a second space probe to steal the show and push back the boundaries of habitability even farther, this time to the confines of the outer solar system. Finally, on July 14, 2015, after a 4.8-billion-kilometer journey, the time had come for New Horizons' long-planned rendezvous with Pluto. A flyby nearly ten years in the making culminated that day with its closest approach. Under the piercing eyes of the spacecraft, the dwarf planet would reveal a completely new and unexpected world.

Traveling at a blistering sixty thousand kilometers per hour, New Horizons gave humanity one of the greatest jaw-dropping moments in recent planetary exploration history. This encounter was a perfect illustration of how our ability to characterize nature relies on our capacity to question it, which depends in part on the technology available at any given time and on our intellectual frameworks. Not even our wildest dreams had prepared us for what we learned. Pluto caught us flat-footed and short of imagination.

When Clyde Tombaugh discovered it on February 18, 1930, Pluto could hardly be separated from the starry background on photographic plates. Only its relative movement gave away its presence. It was found where the astronomer Percival Lowell had predicted the ninth planet would be located, a world so far away that, from its surface, the sun looks just like a bright star in the night sky and shines about as much light on it as the full moon on Earth. It is so distant that everybody predicted that Pluto would be nothing more than just a frozen, dull ice ball drifting at the far end of the solar system. Things stayed that way until its companion moon Charon was discovered on June 22, 1978, by astronomer James Christy. It was confirmed two weeks later after a few follow-up observations, marking a turning point in Pluto's exploration.

With the discovery of Charon, Pluto was not just one out of those many icy trans-Neptunian objects in the Kuiper Belt anymore, a lost region of the solar system located thirty times farther away than the Earth from the sun and a relic of the early solar system's history. Like the main asteroid belt and Jupiter, this donut-shaped region contained enough ice and rocky debris to form another planet. But Neptune's gravity disrupted the area as it moved outward in the solar system. Unfortunately, Neptune's early migration prevented the millions of icy objects it contains to coalesce into a large planet. The Kuiper Belt could also be the reservoir of an estimated trillion or more comets among those objects.

But Charon made Pluto special long before we had any evidence it was unique in its own right. About half the size of Pluto, Charon is the largest moon in the solar system relative to its parent planet, and both orbit around a common center of mass. All of a sudden, we were dealing with a binary system. There was much to be learned from it, which ultimately won the day for the selection of the New Horizons mission years later, and the fact that Pluto lies in a region of the solar system that remained unvisited at the time. The spacecraft would gather data to characterize the formation of Pluto's system, the Kuiper Belt, and the transformation of the early solar system. It would also collect information on Pluto's atmosphere, surface, interior, and environment and its moons before studying other objects in the Kuiper Belt, including the

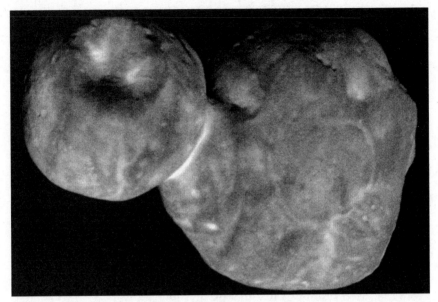

Arrokoth is a thirty-six-kilometer-long frozen world in the Kuiper Belt made of two lobes connected by a narrow neck. Analysis by New Horizons detected methanol, hydrogen cyanide, water ice, and sulfur-rich complex organic compounds on its surface.

thirty-five-kilometer Arrokoth, the farthest object ever approached. By virtue of their incredible distance from the sun, objects like Arrokoth have remained unperturbed over the eons. It showed us what the building blocks of the early solar system looked like and what their composition was, which for Arrokoth was rock, ices, and sulfur-rich organic compounds, including complex macromolecules.

Before the mission took place, the Hubble telescope brought Pluto into greater focus. Although it was still very much a distant and pixelated globe then, the first images covered almost an entire rotation and revealed a complex surface made of regions of substantial differences in brightness. They were interpreted as distinct topographic regions formed by basins and impact craters and by the distribution of frost migrating across the surface with the seasons. Its tenuous atmosphere, one hundred thousand times thinner than Earth's, seemed complex, too.

Made of 98 percent nitrogen, with some methane and carbon monoxide, it expands as Pluto reaches the closest point in its orbit to the sun and collapses as it reaches its farther point. Pluto takes 248 years to complete a highly elliptical and tilted orbit, which takes it within the orbit of Neptune only 4.5 billion kilometers from the sun, or as far as 7.4 billion kilometers away. Its orbit is not its only oddity. Like Venus and Uranus, Pluto exhibits a retrograde rotation and spins on its side with an axial tilt of 120 degrees. Further, patterns of slow twenty-degree oscillations cause ever-changing sets of tropics and an atmosphere that varies in density by a factor of ten thousand over the course of millions of years.

Hubble was essential to discoveries before New Horizons was launched on January 19, 2006. During the spacecraft's cruise, it helped grow the Pluto system to five moons, and its data were used to characterize its dynamics better. Hydra was discovered in 2005 by the Pluto Companion Search Team led by Hal A. Weaver, a group hunting for more and smaller companions of Pluto and Charon ahead of the New Horizons mission; Hydra and Nix were independently discovered by Max J. Mutchler and Andrew J. Steffl the same year; Kerberos was added to the list in 2011 by a team led by my SETI Institute colleague Mark Showalter. One year later, Mark doubled up with Styx on July 11, 2012. Except for Charon, the other moons are at most a few tens of kilometers in size. Kerberos, the smallest, is barely nineteen kilometers in its longest axis. All of them were formed by the coalescence of debris around Pluto after it collided with another Kuiper Belt object, in the same way our moon was formed.

UNEXPECTED ACTIVITY

That was the sum of our knowledge of Pluto's system before the flyby. And then the flyby happened. Among the many iconic images returned by the mission, the blue atmosphere layered in haze bands stays in my mind as one of the most stunning and least expected visions. Ironically, the blue color of Pluto's atmosphere is due to the presence of tholins, the red organic molecules that scatter light at blue wavelengths. As the atmosphere cyclically condenses back to the surface, the complex organic macromolecules fall back and give Pluto the reddish-orange tint observed throughout its landscape. In an opposite process, when the atmosphere expands in summer, part of it escapes into space, and Charon's gravity captures some of it. The tholins fall onto the moon's surface and are responsible for the red spot observed on its north pole.

As Pluto's tenuous atmosphere cyclically escapes into space, it needs a source of nitrogen to resupply it. Like the ocean worlds orbiting the gas giants, the most likely origin for this source would be the planet's interior, but this would require a world considerably more geologically active than anybody suspected before the New Horizons mission. There again, Pluto would not disappoint. Pluto is so far away from the sun that the belief before the mission was that it had started cold. In that scenario, the dwarf planet was an icy world that warmed progressively through the radioactive decay of rocks in its interior. Now the evidence seems to point to a hot start instead and the presence of an early liquid ocean. The "hot start" model refers to how fast Pluto, and other large Kuiper Belt objects, like Eris and Makemake, were formed, probably all of them with liquid oceans. Telltale signs of these early beginnings are still visible today in Pluto's surface morphology. If it had started cold and ice progressively had melted internally, the crust would have contracted over time. As a result, compression features would have formed on its surface. In the hot scenario, the ocean would freeze progressively, resulting in an expansion of the crust and the formation of extension features instead, which is precisely what New Horizons saw, along with a complex geology and an astonishing variety of landscapes that include mountains made of water ice, valleys, plains, craters, polygonal terrain, and glaciers.

Some of these features have become famous since the flyby, like the heart-shaped Sputnik plain. Formed by a 400-kilometer-large impactor 4 billion years ago, the 1,062-by-805-kilometer-wide basin lies 3,000 meters below the surrounding terrain and has accumulated thick layers of nitrogen, methane, and carbon monoxide ices over time. Polygonal patterns

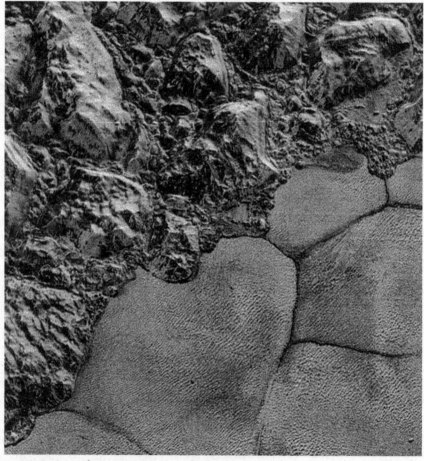

The discovery of diverse mountainous and glacial terrain on Pluto was entirely unexpected. A striking example is Sputnik Planitia's shore, where the al-Idrisi mountains abruptly meet the nitrogen-rich ice plain covered in what resembles convection cells.

cover its northern and central parts. Connected by deep troughs, they could be evidence of convection powered either by nitrogen sublimation or by interior heat from the radioactive decay of rocks. The lack of impact craters on the plain indicates that the process was initiated geologically recently, possibly only 1 or 2 million years ago.

Glaciers made of solid nitrogen ice flow through valleys into adjacent basins and over the hard water ice bedrock from the plain. One of the most striking features of Pluto's surface is probably Tartarus Dorsa, a set of five-hundred-meter-high mountains covered in complex ice-bladed textures looking like spires. Their closest terrestrial equivalents are snow formations created by the combined action of sublimation and wind, also known as "penitents" that appear at high elevation in rarefied atmospheres. Older and fractured terrain dating back from the earliest times of the solar system were also eroded by glacial processes. With Pluto, it seemed that we finally found that special place where hell freezes over.

Charon, in contrast, does not appear to be currently active. The moon is tidally locked to Pluto, orbiting its parent planet in the same amount of time it takes Pluto to complete a rotation in 153 hours. As a result, it always shows the same face to the dwarf planet and never rises or sets. Impact cratering, fractures, faults, and deformation dominate Charon. It is structured around two broad provinces: the northern hemisphere is dotted with cratered terrain, while complex plains cover the southern hemisphere. These two provinces are separated by ridges and canyons created by tectonic processes in Charon's icy crust. The last major resurfacing of the small moon took place 4 billion years ago, and the remaining evidence suggests that it required a source of heat. Charon likely had a subsurface liquid ocean that later froze, explaining the extensional tectonic features observed in its crust. Today, the moon is frozen and too cold to sustain life. Despite the possibility of a subsurface liquid ocean in its distant past, the age of its last resurfacing indicates that habitable conditions did not last. The story might be very different for Pluto.

RELICS FROM AN ANCIENT OCEAN?

Pluto cannot sustain life on its extremely cold surface today but it might still hide a habitable environment if a liquid ocean is present in its deep interior. This scenario is now regarded as increasingly plausible as researchers continue to pour over New Horizons' data. Shockingly enough, for such a small world, we might even have to consider Pluto's past, present, and . . . future habitability.

Pluto's birth would have been rapid and violent with a hot start, its accretion possibly taking place over fewer than thirty thousand years. It could also have happened over a few million years if large impactors buried their energy beneath the surface. In both cases, the energy of rapid accretion and collisions produced enough heat to form a subsurface liquid ocean as early as 4.5 billion years ago. The hot scenario is reasonably well supported by geology on the side of Pluto that was visible to New Horizons during the flyby. Additional evidence was collected on what was visible of the opposite side. The formation of the Sputnik plain by a giant impact 4 billion years ago sent shock waves and stress waves rippling around Pluto, across its surface, and through its center. Energy focused on the opposite point from the impact. It left behind lines in the landscape that look almost like fossil imprints of these waves. Since waves travel at different speeds in different materials, the shape and arrangement of these lines gave away additional information. Pluto would have needed to start with a 150-kilometer-deep ocean and contain the kinds of minerals that form through the interactions between the core and the liquid to fit the observed patterns. The evidence appears reasonably robust. From a habitability standpoint, it implies that Pluto had an abundance of liquid water and energy when Earth was still just a red-hot molten blob of magma. But while water and energy are critical to habitability, they are only part of what makes an environment habitable. Life also needs shelter, energy, nutrients, and sources of carbon.

Assuming it started in the ocean, a habitat located in the deep interior would have provided protection, and the risk for the early ocean to be completely locked in these early times would have been remote. Large impacts created communication pathways with the surface by generating deep faults and fracture systems. Although highly destructive, these impactors had another critical benefit for developing Pluto's habitability and possibly life. Originating in the Kuiper Belt, they were likely ice-rich bodies, including comets, that added more ice and volatiles to Pluto through accretion and delivered abundant organic compounds and carbon. Also, as seen in New Horizons' data, tholins readily form through the interaction of cosmic rays and ultraviolet light with surface and atmospheric methane. As they rained down onto the surface, they could have made their way into the ocean through existing pathways, providing the building blocks of life and nutrients in a top-down (atmosphere to interior) process. The high organics content of the materials that accreted to form Pluto potentially allows a bottom-up mechanism as well (core to ocean). In that scenario, the dwarf planet formed loaded with organic molecules that could have found a favorable environment to jump-start prebiotic chemistry in the ocean and possibly more.

If an ocean still exists on Pluto today, it is the remnant of this early massive body of liquid, probably a mixture of water and ammonia. It is difficult to tell how much of it is left and how much is still liquid. Could we still find some traces of it? We possibly already have. Addressing that question requires us to find evidence of communication pathways that would substantiate its existence and explain how Pluto could have sustained its tenuous atmosphere over time.

For a while, Wright Mons and Piccard Mons, the two highest peaks on Pluto, seemed plausible candidates for such evidence. Both are 150 kilometers across and 4,000 meters high; both exhibit what resembles summit-like depressions and hummocky slopes; and both look like promising cryovolcanoes. The low abundance of impact craters on their slopes and the regional surface hinted at a surface less than a billion years old and probably much younger. But the reevaluation of their topography did not provide the expected high slopes or deep pits that were suspected in New Horizons' high-contrast imagery, in which sunlight hit the two peaks at a high angle. But the disappointment was only short-lived. Since this new topography interpretation, a more detailed geologic mapping of the Wright Mons and its neighboring region revealed more circular mounds. These newly discovered features are possibly formed by the pressurization generated by freezing that creates fractures in the crust and allows liquid to escape from the interior.

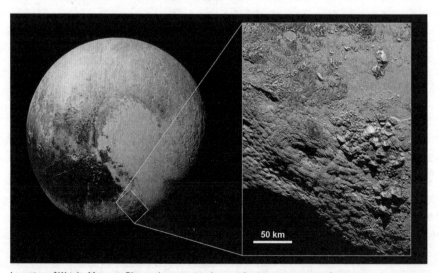

50 km

Location of Wright Mons on Pluto, where regional cryovolcanism is suspected.

This cryovolcanic process could also occur in chambers within the crust or in the subsurface ocean. It may also explain a variety of cryovolcanic formations observed on Pluto. While alternate hypotheses might be proposed in the future, recent findings of short-lived mixed ammonia and water ice deposits added another layer of evidence pointing toward an ocean still lurking at depth. Importantly, this interpretation fits nicely in the big picture of these potential other cryovolcanoes observed on Pluto. The deposits line up with a fractured region named Virgil Fossae, located west of the Sputnik plain, an area of water ice mounds and past tectonic activity. In Pluto's environment, ammonia exposed on the surface is destroyed by the sun's ultraviolet light in less than a billion years. But the time frame can be constrained even further for Virgil Fossae. The absence of impact craters suggests that deposits could be, in fact, less than 10 million years old. The other exciting part is that the pattern looks like a flow from a water ice volcano, and ammonia mixed with the ice is present. Ammonia is a powerful antifreeze, and, in sufficient proportions, it might just be what has kept Pluto's interior ocean from freezing all this time, possibly up until today.

And this might not be the end of Pluto's ocean saga. By the time the sun becomes a red giant over 7 billion years from now, and most of the solar system is all but a memory, Pluto may warm up to reach 27 degrees Celsius. Although this clement period may not last for more than 1 or 2 million years, the dwarf planet that took us all by surprise and forced us to imagine harder, this small ocean world might be the last habitable world standing.

NEW CANDIDATES

Our exploration of the icy moons and the discovery of ocean worlds push back our conceptions of what the last frontier in habitability might be and where life could be waiting to be found. For starters, the Pluto system may only be the tip of the planetary iceberg in the Kuiper Belt, where around thirty-five thousand objects are estimated to be larger than 105 kilometers in diameter. Other candidates could include Makemake (45.8 AU) and Haumea (43 AU), and the list goes on.

With them, we learned that neither size nor distance from the sun should be considered factors to exclude habitability. As a result, some are eager to return to Pluto and Charon for further investigation. The Pluto system revealed an incredible diversity and various features and dynamics that warrant further study. The scientific data acquired by New Horizons and the technological progress made since its launch would give us the ability for much more focused studies and global mapping, including with a subsurface radar sounder.

As a result, a planetary mission concept named "Persephone: A Pluto-System Orbiter and Kuiper Belt Explorer" is currently under study at NASA. Its scientific objectives would be to characterize the internal structures of Pluto and Charon and search for the confirmation of a subsurface ocean on Pluto. It should also investigate the evolution of surfaces and atmospheres in the Pluto system and characterize the evolution of the Kuiper Belt Objects (KBO) population. If ultimately selected, the mission could launch in 2031, make a flyby of the KBO in 2050, and complete a tour of the Pluto system between 2058 and 2061. An extended mission could send the probe back to another encounter with a Kuiper Belt object in 2069.

The flybys of Pluto on July 14, 2015, and that of Arrokoth on January 1, 2019, have left us with some fascinating questions that Persephone proposes to document. A return to this region would help us make scientific leaps in our knowledge of primitive objects, our solar system's early times, and their potential or habitability. For Pluto, unanswered science questions include: Does Pluto have any magnetic field? Are Pluto and Charon fully differentiated? What is the evidence of a subsurface ocean on Pluto? What are the relative ages of, and geological processes acting on, different terrains globally on Pluto and Charon? What is the origin and evolution of Pluto's volatiles? What are the chemical composition and thermal structure of Pluto's and Charon's atmospheres, hazes, and exospheres? What is the composition and escape rate of heavy and light ion species? The study of its moons can also tell us about the origin and evolution of Charon's surface composition and the constraints the small satellites in the Pluto system place on its development. Persephone will also address some critical questions for the KBO, including what can we learn about the collisions in the primordial Kuiper Belt and the formation of its population from the density, shape, and distribution of its objects. What do the surface features of encountered KBO reveal about their origin, evolution, and geologic history? How do their detailed surface properties, composition, and volatiles (and atmospheres, if present) vary?

PUSHING BACK THE LIMITS OF HABITABILITY

While Pluto set an unexpected record for geological activity in such a remote and small world, there might be other frontiers in habitability that have yet to be fully explored. Much closer to home, we have started to question whether even the most unlikely planets or moons could, despite all evidence, have been habitable in the past or even still hide secret oases today.

Only a decade ago, thinking of our moon as a habitable world at any time in its history would have sounded outrageous. Yet, there might be a case for habitable conditions not only once but twice in its early evolution. Not a day goes by without finding more evidence of water on the moon, whether it is water ice in cold traps in permanently shadowed craters, as compositional water in hydrated minerals, or water stolen from the Earth as the moon passes by the tail of the Earth's magnetosphere. The lunar mantle may also contain as much water as Earth's upper mantle.

The moon may have developed a substantial atmosphere just after accretion, and another one a few hundred million years later during its peak volcanic activity. This second atmosphere took about 70 million years to dissipate. It is thus theoretically possible for liquid water to have existed then at the surface and in protected environments in the regolith, the loose rocks and dust that cover it. Whether habitability could have led to life depends on other factors as well, many admittedly still speculative today. One of them is how much time it takes for the building blocks of life to transition to biology. Our only example is the Earth, and from our records, estimates range between 10 million years and as little as a few thousand years.

Even if the building blocks of life were not present on the moon, they could have been delivered by asteroids and comets or transferred from the Earth through impact cratering. Our planet could also have potentially seeded its satellite this way through the transfer of ejected material. Whether that seeding ever took place is unknown, and the early Earth's material transfer to the moon is more than just a plausible theory. There is some concrete evidence that it happened. Among the lunar rocks collected by Apollo 14, one contained traces of minerals with a chemical composition typical on Earth and unusual for the moon. It may have originated from our planet and was blasted off to the moon when the Earth was struck by an asteroid 4 billion years ago. This example of planetary exchange is also what makes astrobiologists consider that, in addition to Mars, the moon might be another good candidate to search for the record of the origins and early days of life on Earth. Early terrestrial rocks blasted to the moon may have been preserved from destruction there when plate tectonics and erosion erased those from our Earth's evidence of deep time.

With the icy worlds of the Kuiper Belt, we crushed the limit of how far from the sun habitable conditions can be found. With the moon, we are now testing the limit of how dead a world may look and still have experienced habitability. And, with Mercury, we may just be pushing the envelope on how close to the sun habitable environments may have once existed. Mercury is so close to the sun that life at its surface would have been simply impossible—ever. However, here is another case where terrain on the opposite side of a planet from massive impact basins may be revealing

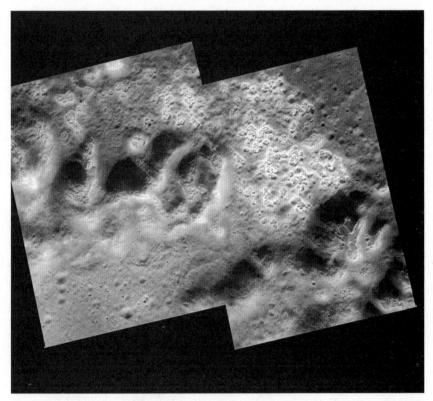

The MESSENGER spacecraft discovered strange depressions on Mercury, the origin of which still remains enigmatic. The size of these hollows varies from 20 meters to 1.5 kilometers in diameter and 40 meters deep.

unforeseen potentials. In the case of Mercury, the region of interest was formed by the massive bolide that excavated one of the largest impact basins in the solar system 3.9 billion years ago: the 1,550-kilometer-diameter Caloris basin. On the opposite side of the planet from Caloris, vast areas of chaotic terrain made of degraded knobby fields reveal the loss of several kilometers of surface elevation. Intermixed with that terrain, ancient lava flows are evidence of a past geothermal disturbance.

This particular type of terrain implies the gradual collapse of volatile elements, possibly through the sublimation of sulfur- and halite-rich materials. Other hypotheses include water, which raises the prospect that some subsurface regions of Mercury could have developed habitable niches over the 2 billion years the process of forming those terrains lasted. Even if habitable conditions persisted underground only briefly, the record of prebiotic chemistry or primitive life could exist in these regions. As astonishing as it sounds, ice is still present today on Mercury in cold traps in those areas that are permanently shadowed. They might be prime candidates for a future visit by an astrobiology mission.

A BRAND-NEW DEAL

Whether the building blocks of life ever assembled or not on any of these worlds that represent the current last frontiers in astrobiology is still unknown. But the more we explore, the more these limits fade away. We have to believe that not all habitable conditions will lead to prebiotic chemistry or life until we are proven wrong. Intuitively, it feels like there could be a gradation between those stages. For now, we are still lacking some of the basic knowledge to understand what the tipping points are. How fast life assembles from its building blocks is probably one of the most critical. The faster it does, the more of what looks today like the last frontiers could, in fact, reveal transient states of living, like ephemeral to long-lasting snapshots in times of our solar system's biological resilience and diversity.

Our generation spearheads a wondrous age of exploration, one of exhilarating discoveries and boundless planetary horizons that endlessly expand our minds. We can pause in wonder, coming to the realization that, although we are still in the process of completing the picture of our own tree of life on Earth, we have started to actively explore a potentially extended family tree inside the solar system. Regardless of how strange new life-forms we might encounter on that journey may be, they will be family to us, literally born under the same sun 4.5 billion years ago, next to us, on our small island in the cosmic ocean. We do not have an answer yet as to how many other beginnings took place, if any, but we can theorize in the light of what we learned so far by exploring the solar system and extreme terrestrial environments.

Theoretically, there could be (or have been) as many unique geneses as habitable worlds. Considering what we know today, that would be a lot of life right there in the solar system itself. But we could be alone as well. The answer to that question could also be anywhere in between, and here, things become very complex very fast. There are many models today trying to shed light on life's origin, but if we want to think about a family tree of life in the solar system, then all of these models can probably be reduced to three scenarios.

One is that life starts on a world and stays confined to it. A second could see life begin on a planet and be exported to another through planetary exchange as impacts of large asteroids and comets blast these seeds of life into space. As this material finds its way to the surface and subsurface of another world, it starts life, provided that it finds a favorable environment. Earth and Mars are strong candidates for this scenario. But life might also have started independently on Earth and Mars and could have made its way in both directions to the other planet. And then there is possibly panspermia, the third-case scenario. Panspermia could have acted as a unique seeding agent throughout the solar system. Its contributing materials, their nature and composition would have depended on the comets and asteroids' origin. Biodiversity on these worlds would have then been dictated by their environments. And those scenarios do not exclude each other.

The answer to the probability and complexity of any family tree of life in the solar system rests in the responses to these questions: Did planetary exchange take place in the inner solar system? We know that the answer to this question is yes, it did. The Martian meteorites found on Earth show that planetary exchange happened in at least one direction. Did it take place in the outer solar system between moons? We do not know yet, but the discovery of fragments of asteroid Vesta on asteroid Bennu shows that this is a universal process that can take place with objects of all sizes. Is panspermia a real engine for seeding life? Some recent discoveries have revived the old concept. It seems like a more real possibility today, but understanding whether it is the primary vector for seeding life in the universe or only an occasional contributor will take more time to figure out. An intriguing question that remains as well is whether those processes could have worked in concert, in which case, the solar system's biological heritage might represent a much closer family than we think. Finding the first biosignatures outside Earth in the solar system will shed some light on these questions, most of which can also be extended to any other planetary systems beyond ours.

8

REVOLUTIONS IN THE NIGHT SKY

Hardly a day has gone by in the past decades without producing the discovery of a new faraway world. We now blissfully take it for granted that other planets exist beyond our solar system, revolving around alien suns or other strange stars in the night sky. It takes an exoplanet to be especially weird, or to resemble the Earth, to make it to the front page of the news nowadays. Exoplanets rarely appear in the world of headlines lately, a clear sign that they are now integrated into our universal consciousness in a sort of business-as-usual kind of way.

TOWARD A COSMIC PLURALISM

Not so long ago, in the grand scheme of humanity's history, Giordano Bruno was burned at the stake as a heretic in the opening months of seventeenth-century Italy, in part for daring to think that there might be other inhabited worlds in the universe. Pushing the heresy further, he also suggested that some other worlds could foster life. He was not the first one, either. The plurality of worlds, the belief that numerous worlds exist in the universe and harbor life, extends back to antiquity with Thales in Greece and Lucretius during the Roman Empire. In contrast, Plato and Aristotle viewed our planet as unique. Mainstream Christian doctrines subsequently echoed this idea. There were notable exceptions in theologians who believed that the existence of extraterrestrial life was consistent with the Christian faith, but with the dominant influence of the church in Europe, the idea of multiple inhabited worlds was set aside for another thousand years, at least officially. Some Muslim scholars endorsed the notion of cosmic pluralism during medieval times as well. Buddhism holds that there are countless worlds inhabited by sentient beings, although not necessarily in this universe. Meanwhile, Hinduism always taught the concept of multiverses and multiple dimensions in a vast cosmic ocean of innumerable planetary systems where alien life abounds.

It would take the Copernican revolution and the invention of the telescope to start shaking the status quo. The Ptolemaic model of a stationary Earth at the center of the universe was replaced by the Copernican heliocentrism, which placed the sun at the center of the solar system. The planets revolving around it were observed closer than ever before with the newly invented telescope. Then these "wandering stars" of the Greeks emerged for the first time in the field of view as real worlds in their own rights. Astronomers started to describe their individual physical characteristics and cycles. Newton's mathematically demonstrable universal laws seemed to naturally open to cosmic pluralism as a statistical outcome in the century that followed. At the same time, his followers made the belief in intelligent extraterrestrial life an orthodox component of Newtonianism.[1]

The debate goes on today in multidisciplinary forums between scientists, philosophers, and theologians, each enriching the conversation with their own perspectives. With the mounting evidence of how common the building blocks of life are and the thousands of exoplanets already detected in just a small corner of our galaxy, the more outlandish idea is that life could be unique to Earth. Today, theology appears ready to take on cosmic pluralism in a modern vision of religious faith in which aliens are part of God's plan, not in contrast with it. This concept was already central to Angelo Secchi's view of the universe as he discussed the existence of other inhabited worlds he was convinced existed. A Jesuit priest, Secchi (1818–1878) was one of the founders of modern astrophysics and director of the Astronomical Observatory of the Collegio Romano in Italy.[2] In the twenty-first-century world, we have come to a point where our civilization is on the verge of becoming interplanetary. Every single day, scientific discoveries show us that our place in the universe is not special. Yet, we have emerged from it after 4 billion years of evolution as living conscious beings. For thousands of years, we thought about a cosmic fellowship that may be drifting as we are in the vast expanses of the universe. We fantasized about whether alien worlds could exist or were just a figment of our imagination. We wondered about their mysterious landscapes and whether they would be inhabited. These fantasies fed generations of essays, science fiction novels, B movies, and Oscar-worthy films. What is very different now is that the evidence is overwhelming: these worlds do exist, and for the past thirty years, we have become increasingly familiar with them and, even more so, with their staggering abundance. They gave us visions of Tatooine and Mordor and, in their similarities and differences, have taught us what our own beginnings as a planetary system might have looked like 4.6 billion years ago. They also gave us a tangible basis to think about what type of life could evolve on them.

Now an integral and essential part of the search for life beyond Earth, the hunt for exoplanets is not without its challenges. The first one is fairly obvious: Planets are not stars. They can be bright or dark or anywhere in between, but they can only reflect the light of their stars, and, depending on how close their orbit takes them, they may become lost in their light. Even the biggest planet is usually relatively small. Because of distance and size, it may be only a fraction of a pixel or a few pixels at most in an image. To make matters more complicated, all of these issues are often compounded. As a result, it takes patience, resolve, and repetition to detect a candidate exoplanet. Several methods are available to separate the elusive planetary dots from the stellar background in this cosmic hunt. More than one can be used to study a candidate and confirm it.

DOTS IN THE NIGHT, OR HOW TO DETECT EXOPLANETS

A star wobbles ever so slightly in response to gravitational tugs when planets are present. This wobble translates into color shifting in the star's natural spectrum, making it appear bluer or redder due to the Doppler

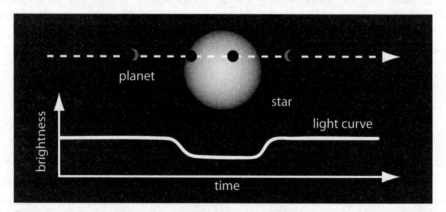

The transit method involves studying the brightness changes of a star as a planet passes in front of it. Astronomers use this method to calculate a planet's size by analyzing the brightness curve and knowing the star's dimensions. They also measure the planet's orbital period through the time between transits. With the period known, Kepler's third law is applied to find the planet-star distance. From it, scientists can start modeling planetary environments.

effect. This way of tracking exoplanets is the radial velocity method. Another one is to search for shadows. That is the transit method. When a planet passes directly between its star and the observer, the star's light dims ever so slightly, and that dip in light intensity is measurable. In other words, this is a photometric method. For instance, transits by terrestrial planets may dim a star's light by 1/10,000, which speaks about the precision required to find these alien worlds.

Direct imaging is now another method available with increasingly more powerful telescopes and imaging capabilities. In their case, astronomers suppress the star's glare by masking it, and they search for the light reflected from a planet's atmosphere in the infrared. They also apply Einstein's theory of general relativity and use the gravitational force of distant objects to bend and focus light coming from a star. This indirect method is known as gravitational microlensing. As Einstein showed, gravity warps space and time. A large mass like a star causes the fabric of space to bend around it, making gravity distort and focus light, like the lens of a magnifying glass. Gravitational microlensing enables astronomers to find planets using light from a distant star when its path is altered by the presence of a massive lens, in our case, a star and a planet. As a result, the distant star will appear brighter for a short period of time. This method requires the observation of many stars over long periods, since no one can predict when or where microlensing will take place. The last method is astrometry, which detects planets around stars by measuring minute changes in the star's position as it wobbles around the center of mass of the planetary system.

The transit method accounts for about 75 percent of exoplanets detected so far, mainly because it was used by the Kepler and K2 missions, and now by TESS (the Transiting Exoplanet Survey Satellite), two space telescopes dedicated to exoplanet hunting in recent years. Although these numbers change almost daily, over 5,573+ exoplanets and 4,151 planetary systems have now been confirmed in just a very small quadrant of our galaxy. Close to 10,085 additional exoplanets still await confirmation. If we infer from these numbers, there could be at least one planet for every star, thus 100 to 400 billion in our galaxy alone, and probably many more. Admittedly, not all of them will be in the habitable zone of their parent stars. However, exploring our solar system taught us that an object does not need to be a planet or within the habitable zone to develop habitable environments. Assuming that the solar system is relatively representative of the ratio of moons to planets, there could also be an estimated 20 to 80 *trillion* moons in our galaxy alone—and the Hubble Deep Field image suggests the existence of 125 billion galaxies in the observable universe! The math is simply staggering. As a scientist, I will go out on a limb here and say that if life is an accident, the universe has to be the mother of all pileups. In contrast, although predicted, moons are elusive at this point, and those suspected still await confirmation. Tomorrow, with increasingly sensitive technological means of detection, they will flood the statistics. This is simply a case where our understanding of the abundance, distribution, and size of exoplanets and exomoons is tied to the resolution of our instruments and the effectiveness of our methods to find them.

Early in the search, hot Jupiters seemed to overwhelmingly populate the night sky. These are planets around the mass of Jupiter but orbiting much closer to their stars (about 0.1 AU). In reality, their size made them the easiest objects to detect with our current technological means. As time went on and technology improved, we learned to search better, too. As a result, smaller objects have been discovered, potentially including the first large exomoon eight thousand light-years away, Kepler-1625b-i, announced on May 10, 2016. Approximately the mass of Neptune, it could orbit a parent planet several Jupiter masses in size. More recently, in 2022, another candidate was announced,

Kepler-1708b-i, a possible exomoon 2.6 times the size of the Earth. The size of these moons is disproportionate compared to anything we know in the solar system, a size bias that could be due to the resolution of the detection method. The search for exomoons is in great part performed using the transit method, so it is difficult to separate their signal from their parent planets. But with the new generation of telescopes, whether ground-based or spaceborne, an increasing number of smaller objects will be detected.

Kepler and TESS have shown the way in the past thirteen years. The James Webb Space Telescope (JWST) has joined the search, giving us the first confirmation of the presence of a greenhouse gas (CO_2) in WASP-39b's atmosphere. In the next few years, we may have a more accurate

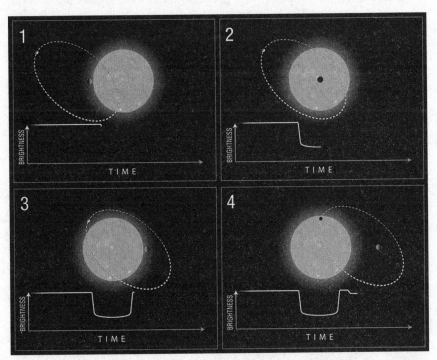

The Hubble telescope recorded the transit of Kepler-1625b-i, causing periodic dips in star brightness. A second dip hints at a possible exomoon, similar in size to Neptune. Further observations are needed to confirm this discovery.

representation of the size/frequency distribution of the various types of exoplanets and exomoons in our galaxy. But regardless of our limitations, we ought to stop and marvel at how far we have come in only a few decades. It is hard to imagine that the modern era of exoplanet detection just started in the 1980s. The pace of discoveries has only accelerating since, and it began with ground-based telescopes.

FIRST DETECTIONS

For many reasons, the Atacama Desert in Chile is intimately connected to astrobiology, and its exploration helps us to prepare planetary missions searching for biosignatures, particularly in analog environments to early Mars. But its connection is as deep with astronomy—which won't surprise anybody considering the night sky there. Among all the observatories established in the Atacama, Las Campanas stands on a long ridge overlooking the desert. Today, it is the site of the twin 6.5-meter Magellan Telescopes on Cerro Manqui, which will soon be joined by the Giant Magellan Telescope (GMT). It will be operational in 2029 and will explore the distant universe searching for biosignatures on exoplanets. Las Campanas is also the location of the 2.5-meter du Pont telescope, where the instrument captured an image that made headlines worldwide in April 1984. It showed a disk of dust around the star Beta Pictoris. The gap in the disk suggested that planets might have formed there, but astronomers could not detect them at the time.

The first suspected detection of an exoplanet occurred in 1989 by a team led by astronomer David Latham, who identified an object orbiting a star 126 light-years away in the constellation Coma Berenices. Thought to be a brown dwarf at first, HD 114762 b was confirmed as an exoplanet only thirteen years later. After its launch in 1990, the Hubble Space

Telescope (HST) began supporting the search for exoplanets, among its other observations. In 2001, using data from HST, David Charbonneau and Timothy Brown made the first detection of an exoplanet's atmosphere and analyzed its composition. The planet orbited a sun-like star located 150 light-years away from Earth in the constellation Pegasus. By then, we already had proof that alien worlds lurked in the night sky. In 1992, the astronomers Aleksander Wolszczan and Dale Frail were the first to provide evidence of the existence of exoplanets around a pulsar 2,300 light-years away in the constellation Virgo. The star should have pulsed every 0.006219 seconds, but was a little off from time to time. Intriguingly, the offbeats came at regular intervals. Hidden in these signals were two planets, one three times and the other four times the mass of the Earth. These were the first two exoplanets detected and confirmed but, considering their extreme radiation environment, none of them were likely to harbor life.

The first exoplanet orbiting a main-sequence, sun-like star, 51 Pegasi b, was discovered three years later, on October 6, 1995, by Didier Queloz and Michel Mayor at the Observatoire de Haute-Provence in France. Half the size of Jupiter, it had an orbit that took it exceptionally close to its star. Many more similar exoplanets would be discovered in the following decades and would become commonly known as hot Jupiters. And the discoveries kept on coming. Another exoplanet orbiting a sun-like star (HD 209458) was discovered independently in the constellation Pegasus by two teams led by David Charbonneau and Greg Henry in 1999. It was soon followed the same year by the first multi-planetary system known to orbit a main-sequence star, also the first system known to orbit a multiple-star system. At the time of the discovery, only two planets were detected. Two more were identified later, all four the size of Jupiter. Then a planet was found for the first time within the habitable zone of its star on April 4, 2001. The discovery was made by a team of Swiss astronomers led by Nuno Santos from La Silla, another Chilean observatory in the Atacama Desert. Located 128.6 light-years away from Earth in the constellation Eridanus, the Jupiter-sized HD 28185 b orbits nearly at the same distance as the Earth from the sun. As the discoveries steadily came in with various detection methods from ground-based instruments, the nascent era of space telescopes was about to revolutionize what we knew of our night sky. Launched within a few years of each other, these space observatories would provide powerful and complementary detection tools and light up the galaxy with heaps of alien worlds.

OBSERVATORIES IN THE SKY

On June 22, 2003, the Canadian Space Agency launched the MOST (Microvariability and Oscillations of STars) mission. The 143-pound satellite carried a high-precision telescope, no larger than a plate, capable of measuring light oscillations in stars. It was the first spacecraft dedicated to asteroseismology. It was also capable of measuring reflected light from exoplanets and detecting them using the transit method. The Spitzer Space Telescope joined the search two months later, collecting data about exoplanet sizes and atmospheres. Less than two years after the beginning of the mission, astronomers using Spitzer announced the direct observation of infrared light from a couple of hot Jupiters for the first time, HD 209458 b and TrES-1b.

In a time of firsts, HD 209458 b represented a milestone in many ways. This exoplanet, located 159 light-years away, is the closest hot Jupiter to Earth and a gas giant orbiting only 7 million kilometers from a sun-like star. It was discovered from the ground first in 1999 and then monitored from space. This exoplanet was the first known to have an evaporating hydrogen atmosphere containing carbon, oxygen, and water vapor. And HD 209458 b was the first extrasolar gas giant to have a super-storm measured. Spitzer data was also used in May 2007 by astronomers David Charbonneau and Heather Knutson to produce the first map of an exoplanet. The map showed the temperature of the cloud cover on HD 189733 b, located 64.5 light-years away in the constellation Vulpecula (the Little Fox). HD 189733 b and HD 209458 b were the first two exoplanets to have their atmosphere analyzed by Spitzer through spectroscopy, a promising method to detect biosignatures.

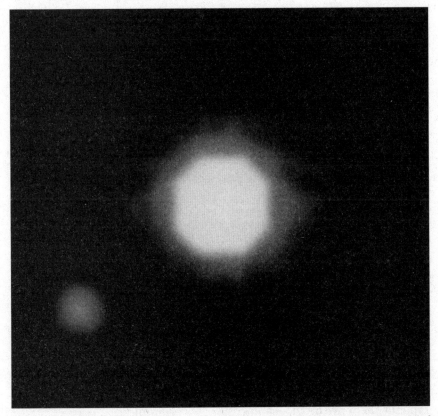

The European Southern Observatory's Very Large Telescope (VLT) captured the first-ever image of an exoplanet, 2M1207b, in 2004. This exoplanet, located 230 light-years in the Hydra constellation, orbits a brown dwarf, making it the first direct image of an exoplanet and the first found around a brown dwarf. Confirmed in 2005, 2M1207b is a Jupiter-like planet, five times more massive, located twice the distance from its star as Neptune is from the sun.

In 2006, the National Center for Space Studies (CNES), ESA, and international partners launched CoRoT, the Convection, Rotation et Transits planétaires space telescope, which operated between 2006 and 2013. CoRoT focused on searching for exoplanets with short orbital periods and performed asteroseismology by measuring solar-like oscillations in stars. It made its first discovery within a year of reaching space. Later in 2009, it found the first exoplanet with a rock or metal-dominated composition. Over the six years of its mission, CoRoT surveyed over 150,000 stars and discovered thirty-four exoplanets.

A REVOLUTION NAMED KEPLER

A little over a decade after confirming the first alien world, three types of exoplanets were known to astronomers. They were the gas giants, the hot super-Earths in short-period orbits (exoplanets larger and more massive than Earth, but less massive than Neptune), and the ice giants. Lacking from this list were terrestrial planets—worlds up to twice the size of Earth. Finding them was one of the primary objectives of Kepler, a mission launched on March 6, 2009. Kepler was, without any doubt, a transformational moment in astronomy and space exploration. It represented a paradigm shift in how we perceived the night sky and completely changed our understanding of our place in the universe. Before it was launched, we already knew exoplanets existed and had confirmation of a few. However, as soon as Kepler started to operate, the sky filled up with worlds we could not even have imagined existed, and the spacecraft brought them by the thousands.

Kepler is also the story of the mission that almost never was, if it had not been for the resolve of one man, William "Bill" Borucki. Bill was a space scientist at NASA Ames Research Center, now retired, and the principal investigator of the Kepler mission. I was fortunate enough to cross paths with him many times at NASA Ames and at the SETI Institute, where the ties with Kepler run deep. Bill is a gentle, soft-spoken, humble, and outstanding scientist with a special grace about him. I never saw him impatient, a quality that probably saved Kepler. Developed over several decades, the mission concept was designed to study the abundance of Earth-sized planets in the habitable zone of sun-like stars in our galaxy. Bill knew the limitations of ground-based astronomy for this type of exoplanet and thought their detection would be best accomplished from space. Between 1984 and 1987, as the idea gained momentum, NASA Headquarters financed proof-of-concept instruments. In 1992, a first mission concept was proposed (FRESIP: FRequency of Earth-Size Inner Planets), but it was rejected. Between 1992 and 1999, the proposal would be resubmitted and rejected four times before finally being selected in December 2001 as a NASA Discovery mission under the name Kepler. Team members Dave Koch, Jill Tarter, and Carl Sagan requested the name change to honor Johannes Kepler (1571–1630) and the laws of planetary motion he defined. From then, another eight and a half years would pass before the space telescope was ready to launch on March 7, 2009. And from its observation one month later to the moment it was finally retired on October 30, 2018, Kepler rewrote astronomy.

The Kepler mission gave us another "overview effect" moment. This time, we were not looking back at our planet from space anymore. Instead, we found our place in space, peering outward into the universe. And the message Kepler sent us was that we were one out of many. Over 530,500 stars were observed during the mission, and 2,662 exoplanets were confirmed. Of those, some were near Earth-sized planets in the habitable zone of their stars. Many more exoplanets are still awaiting confirmation in the mission's archives.

This mission alone produced close to three thousand scientific publications. These results are mind-blowing in more ways than we can think of. This avalanche of new worlds was discovered in a very small patch of our galaxy in just the constellations Cygnus and Lyra. Out of the 4.5 million stars in the telescope's field of view, Kepler monitored over 170,000. Also remarkable is the fact that these results were achieved even though the primary mission ended prematurely due to the loss of the telescope's second reaction wheel on May 14, 2013. Undeterred, the Kepler team found a way to continue operating it using the pressure of sunlight to stabilize the telescope. In June 2014, thanks to lots of ingenuity and some physics, the K2 mission was born. It would open new areas of the sky to the telescope until Kepler ran out of fuel and was retired.

EXHILARATING NEW MISSIONS

TESS was launched almost nine years to the day after Kepler's first observation. This NASA Astrophysics Explorers mission picked up where Kepler left off to find potential planets orbiting bright host stars relatively close to Earth using the transit method, with approximately five hundred thousand stars to be studied. Conversely to Kepler, TESS is not confined to observing a small quadrant of our galaxy. Instead, the telescope covers 85 percent of the sky. Over the next two decades, one of its main objectives is to catalog thousands of exoplanets, including hundreds smaller than the Earth. This survey should reduce the size distribution bias of early detections of large-sized planets and give us a more accurate vision of the abundance of these worlds, from small rocky worlds to giant planets.

Beyond its own mission's goals, TESS is part of a larger exoplanet exploration arc. Because TESS targets relatively bright stars, discoveries can be followed up with ground-based spectroscopy to measure the exoplanet mass and with spaceborne spectroscopy to characterize their atmosphere. The mission also measures the size of planets. With both mass and size, density can be inferred, and these measurements allow the modeling of what exoplanets are made of.

They also provide a foundation for other missions, such as ESA's CHEOPS (CHaracterising ExOPlanet Satellite), which launched a little over a year after TESS, on December 19, 2019. CHEOPS is the first mission designed to follow up exoplanet discoveries by measuring the size of known transiting exoplanets orbiting bright and near stars and searching for predicted transits of exoplanets previously discovered via radial velocity. These well-characterized worlds can then be targeted by the James Webb Space Telescope (JWST) in space or by ground-based telescopes, such as the Extremely Large Telescope (ELT) located in Chile on Cerro Armazones in the Antofagasta region, which is part of the European Southern Observatory (ESO).

JWST is the newcomer. After over twenty-five years of development, significant delays, a redesign in 2005, and a pandemic, the telescope was finally launched on an Ariane 5 rocket from Kourou, French Guiana, on December 25, 2021. It reached its permanent residence, a million kilometers away from Earth, almost a month later. The telescope adds a pivotal piece to the international network of telescopes dedicated to exoplanet exploration in space and on Earth. Thanks to the precision of the launch, the space observatory should have enough propellant to support science operations for much longer than the original ten-year science lifetime. It will possibly even double it.

The most ambitious and complex space telescope ever built is an international collaboration between NASA, ESA, and the Canadian Space Agency (CSA). One of its primary objectives is to shed light on our cosmic origins by observing the first galaxies and the birth of stars and planets. Nothing short of a quantum leap in astronomy and cosmology is expected from it. JWST will also sound the depths of space in search of exoplanets with a potential for life, performing the first detailed near-infrared studies of the atmosphere of habitable-zone planets.

The Nancy Grace Roman Space Telescope (formerly known as WFIRST) is currently under development and should join the hunt no later than 2027. The mission will have multiple astrophysics goals beyond completing a census of exoplanets using gravitational microlensing. This survey could help detect exoplanets down to a mass only a few times that of the moon. It will also look for rogue planets, worlds that are free-floating in interstellar space without parent stars, down to the mass of Mars. The Roman telescope will carry a coronagraph for exoplanet imaging to produce the first direct images and spectra of exoplanets resembling our own gas giant planets.

We have a lot to look forward to, especially since advances are not only taking place in space. This revolution in astronomy is very literally mirrored on the ground by extraordinary innovations in technologies, instruments, techniques, and exploration methods. The atmospheres of exoplanets can now be revealed through optical interferometry, the combination of signals from two or more telescopes to obtain measurements with higher resolution than with only one, while advanced spectroscopic life-detection methods are being designed. These techniques, emerging in an era of giant telescopes, segmented mirrors, and adaptive optics, will make many upcoming ground-based telescopes more powerful than current space telescopes. The Large Synoptic Survey Telescope will detect a wide range of exoplanet populations; the Gemini Planet Imager is already imaging them, and with a resolving power ten times greater than the Hubble Space Telescope, the Giant Magellan Telescope will measure their velocities and atmospheres.[3] Exoplanet detection is a highly dynamic field that is likely to remain vibrant and exciting for many decades. It is rewriting what we thought we knew about the range of possible planetary environments. The search also teaches us how we came to be as a planetary system and breaks open the field of possibilities for life as we know it and life as we do not know it.

9

VISIONS OF TATOOINE AND MORDOR

The alien worlds we have discovered are so far away that the method to categorize them, for now, is through their diameter and mass. Soon, thanks to the new generation of space and ground-based telescopes, we will know more about their atmospheres. In the meantime, using these criteria, the 5,573+ confirmed exoplanets are classified either as gas giants, super-Earths, Neptune-like (Neptunians), Hycean worlds, or terrestrial (rocky) planets.

AN INCREASINGLY LARGE GALLERY OF NEW WORLDS

Rocky planets are worlds about the size of Earth or smaller. Super-Earths, also called mini-Neptunes, range between 1.5 and 10 Earth masses. Their name only refers to their size and does not have any implications for their habitability. They are intriguing worlds found in abundance among exoplanets and have no tangible equivalent in our solar system. They can be scaled-down versions of our gas giants or scaled-up versions of our terrestrial planets—or something entirely different, as their character-istics vary as well. Some of them are mainly composed of hydrogen and helium, while others with higher density are water-rich or silica-rich. A general rule of thumb appears to be that density increases with diameter up to two Earth radii. For the larger ones, density drops rapidly after that. This observation tells us that super-Earths around 1.5 Earth radii are likely to be ocean planets or rocky planets with a thin atmosphere.

Hycean worlds fit between the super-Earths and Neptunians. Al-though they do not resemble Earth-like planets, they could still be hos-pitable to life. They appear to be abundant and can be up to 2.5 times larger than Earth, with vast oceans of liquid water, hydrogen-rich atmo-spheres, and an environment that could host extremophiles. They might become outstanding candidates to search for biosignatures in exoplanet atmospheres. The larger Neptunian exoplanets are comparable in size to Neptune or Uranus (about fifty thousand kilometers). They may have rocky interiors, hydrogen and helium atmospheres, and would be consid-ered ice giants in our solar system. JWST will peer through their thick layers of clouds and give us a look at their spectral signatures. Rare cases of "hot Neptunes" have been detected, too, but they might be planets in transition, possibly Neptunian-sized worlds that will ultimately lose their atmosphere and become super-Earths.

Gas giants exceed ten Earth masses. They may have a rocky core but are primarily composed of hydrogen and helium. Hot Jupiters or-biting 0.015 to 0.5 AU from their stars represent a subcategory of these exoplanets. From their composition, it is easy to see why they should not be found where they are, at least in theory. In our solar system, gas giants form beyond what is called the "snow line," a region where water and ammonia are frozen. Therefore, the current location of Hot Jupiters suggests that they started beyond the snow line and then moved inward toward their stars.

Our own Jupiter undertook such a migration in its early days. After moving inward to 1.5 AU (about the current position of Mars today), it reversed course and migrated outward, pulled back by the formation of Saturn. If Jupiter's migration had not been halted, it would have become a hot Jupiter. But hot Jupiters might not all start the same way. A new theory shows that they may have formed close to their suns, where they are observed, without evaporating away. Typically, if they existed in our solar system, these planets would be located within the orbit of Mercury or even closer to the sun. They start in a process called core accretion in the protoplanetary disk that leads to the formation of a rocky core the size of the Earth or larger. As the core reaches ten Earth masses or more, it undergoes extremely rapid accretion, which pulls enough hydrogen and helium into the atmosphere for the planet to become a gas giant. This new theory stems from the observation that, instead of being scattered randomly, hot Jupiters have a relatively sharp inner boundary in their spatial distribution that lies within specific radii of their stars. Their core could be inherited from their beginnings as super-Earths, and it would take only 1 percent of those to undergo a runaway accretion to explain the existing population of hot Jupiters. If this new theory is correct, and observations appear to support it, most could have formed where we see them today. It would also mean a distinct origin for our own Jupiter and raises an intriguing question about a possible relationship between the absence of super-Earths in our solar system and the lack of hot Jupiters.

In the Grand Tack theory, as Jupiter migrated inward toward the sun, it destroyed an early generation of planets before moving back outward to its current position. While this resulted in the subsequent formation of Mercury, Venus, and Earth, as well as the stunted development of Mars, it could also signal that planetary systems like ours are not that common. Indeed, as we look at their structure and composition so far, very few look anything like ours, and one may wonder whether this could have any implications for the emergence of life. In the meantime, we must start with what we know that worked for life on Earth. Then we will see how our own equation of life can be adapted to the various types of exoplanets and where they are located in the habitable zone of their parent stars. In that perspective, the ever-growing number of identified planetary systems already helps us refine our definition of habitability and reexamine the number of potentially habitable worlds within our galaxy. Here, too, the Kepler mission was foundational in giving us a new perspective.

Before Kepler, we knew of about a few dozen exoplanets, most of them extreme worlds by any stretch of the imagination. Since then, Kepler showed that there could be 300 million potentially habitable worlds in our galaxy alone in the most conservative estimate, including a few within twenty to thirty light-years of the Earth. This sheer number is reached by considering that only 7 percent of sun-like stars could be hosts of habitable worlds, but this total could have been substantially underestimated, and its current reevaluation is not the result of a sudden excess of optimism, either. A new evaluation was based on a recent study focused on rocky worlds 0.5 to 1.5 times the size of Earth orbiting sun-like stars, with similar age and temperatures as our sun (within plus or minus 850 degrees Celsius). It analyzed the effect of the type of light given off by stars in this range and absorbed by a planet. Combining NASA's Kepler data on exoplanets and ESA's Gaia mission data on the amount of energy delivered to a planet opened an entirely new horizon for habitability. We could now be looking at 50 to 75 percent of sun-like stars with rocky planets capable of sustaining liquid water on their surface, representing roughly 2 to 3 billion planets! Of course, water is just one of the critical elements for life, but the larger the number of habitable worlds we start with, the greater the chances other living worlds are waiting to be found.

Exoplanets give us a new gallery of alien worlds to think about. As an astrobiologist, this is what I find the most exciting about them because they open up countless unexplored probabilities of the coevolution of life and planetary environments. In terms of habitability, they go from the hellish to the promising and from the slightly unusual to the utterly improbable. And, for each of them, we must evaluate whether they meet the necessary and sufficient conditions for the emergence of life, acknowledging that we are still greatly data-limited in our evaluation. As a result, just as when we explore habitability and life in our solar system, we start with life as we know it. Then we can move on to the unfamiliar territories of life we may not know, where some of these worlds will undoubtedly take us. Here, new analytical and modeling tools supported by artificial intelligence (AI) and machine learning (ML) can help us tremendously.

ROGUE PLANETS

Among these newfound worlds, either individually or as planetary systems, some completely depart from anything we know, whereas others remind us of slightly different versions of home. Of the most extreme and surprising exoplanet findings are rogue planets, which have to be very high on the list of strange things. These lonely worlds wandering in interstellar space are formed from various processes. They have been either ejected from their parent systems through gravitational tugs during planetary migrations or as the result of failed stars that never ignited and are known as brown dwarfs. There could be billions of these interstellar travelers roaming just in our galaxy. The first one to be detected was CFBDSIR2149, a rogue planet discovered in November 2012 during joint observations between the Very Large Telescope (VLT) in Chile and the Canada-France-Hawaii Telescope (CFHT) in Hawaii. This nomad world was caught wandering about one hundred light-years away from us and could be 50 to 120 million years old. Many more rogues followed. Recently, their number almost doubled when astronomers discovered between 70 and 170 of them (some await confirmation).

With a mass ranging from four to thirteen Jupiters, and located 420 light-years away, this particular group was detected through direct imaging in the visible and near-infrared. These rogue planets are very young (3 to 10 million years old) and still hot enough to make their detection possible with this technique. Considering that rogues are orphan planets without parent stars, it would seem almost incongruous to even mention their habitability potential since the notion of a habitable zone is usually associated with stars—a range of distances to a star where water can be liquid at the surface of a planet. But that would be forgetting the lessons taught by our own solar system, where, beyond the habitable zone, moons and planets can still harbor habitable environments. Yet, can we seriously consider that rogues could push the notion of habitability to such extremes? And what about life? As surprising as this may sound, this might not be entirely impossible.

Many rogue planets detected so far are still hot from their recent accretion process, which explains why they can be detected in the near-infrared. Their young age is inferred based on surface temperature modeling. If a planet is flung out from the orbit around its star early, it cools down in outer space. While a thick blanket of ice might help prevent an ocean from completely freezing, radioactive elements deep in the core also contribute to keeping water in its liquid state by warming the planet from the inside. It would seem then that the earlier rogue planets are ejected, the hotter they are, and the best chance a subsurface ocean stands to last. And while water is only one element of the equation of life, planets come loaded with the building blocks of life.

As a result, even planets lost in space could harbor an inner ocean, providing an environment for prebiotic chemistry and biology, a shelter for life, and an energy source produced by geothermal heat. Meanwhile, primitive organisms could extract nutrients from minerals containing sulfur and iron around hydrothermal vents at the contact of the hot rock interior and the ocean. Perhaps one of the key questions for rogue planets is understanding how fast life can emerge. If it does develop before the planet is flung out of orbit, it could stand a chance to survive for some time after the planet starts to wander. The duration of biological processes would depend on the size and type of the planet, how fast it cools down, and the presence of an ocean.

There might be other scenarios, including one that allows rogue planets with extremely dense hydrogen atmospheres to trap enough heat to keep surface water from freezing. In theory, this type of environment could also support primitive life. Or, a best-case scenario might be for a rogue planet to have a companion moon big enough to generate tidal heating. This scenario takes us back to a Jupiter-Europa or Saturn-Enceladus dynamic, where the moon's interior stays warm and develops an interior ocean instead of the planet. Models generated by Manasvi Lingam and Avi Loeb, who theorized about the potential for life on rogue planets, suggest that photosynthesis could still be enabled on rogue worlds located within one thousand light-years of galactic cores in galaxies with an active nucleus.

In the coming decade, more observatories in space and on the ground will focus on studying these transient objects. They will monitor asteroids and comets not bound to a star that may pass through our solar system, like 'Oumuamua or 2l/Borisov did recently. These wandering worlds are another reason the theory of panspermia is being revisited these days, as they may provide a possible way for life to jump from one planetary system to another. Mission concepts to these interstellar objects are germinating as well. However, unless they enter our solar system, our technology is not quite ready yet to reach them and sample them.

On the other hand, if they enter the solar system, there might be ways to take advantage of their visit. For example, NASA is working on a mission called Extrasolar Object Interceptor and Sample Return. ESA plans to explore these objects with a mission called Comet Interceptor. It is hard to predict when one of them will swing by our planetary neighborhood next. Because of this uncertainty, these missions would be sent

'Oumuamua, a Hawaiian term for "messenger" or "visitor," was the first known interstellar object to enter our solar system, and its true nature is still debated. Top: Artistic representation (source: M. Kornmesser/ESO). Bottom: Image of 'Oumuamua (circled in the center) obtained by combining several images from the VLT and the Gemini South telescope.

ahead of time and parked, waiting for these interstellar visitors to show up. ESA will direct its spacecraft toward the sun-Earth L2 Lagrange point and leave it there. Meanwhile, NASA will send a probe toward Jupiter and wait for an interstellar object to enter our interplanetary space before activating the spacecraft for a rendezvous and a sample return mission.

WHEN REALITY MEETS FICTION

While rogue planets represent unbound interstellar vagabonds, other exoplanets closely tied to their stars offer a wide range of planetary environments. They go from the utterly inhospitable to those coming close to representing a potential Earth 2.0. If we consider the most extreme and unhospitable worlds discovered so far, 55 Cancri e is probably very close to the top of the list. It is a super-Earth located forty-one light-years away from us and embodies its own planetary version of Dante's inferno. Discovered in 2004, it is about twice the size of the Earth and eight times more massive. Orbiting twenty-six times closer to a sun-like star than Mercury, it takes less than eighteen hours to complete an orbit around its parent star. 55 Cancri e could be composed of light elements and compounds. It is so hot ($2,700°C$) that it contributes to a supercritical fluid state where gases behave almost like liquids. Its surface is probably made of lava. Its atmosphere is composed of hydrogen and helium, with silicates condensing into clouds in a gloomy, sparkling atmosphere.

But 55 Cancri e might only represent a mild version of another scorcher planet. This one is a hot Jupiter as dark as coal and so close to its parent star that it only takes a little over a day to complete a revolution. Imagine a gas giant one and a half times the size of Jupiter zooming around the sun in just one day. That is the description of WASP-12b and why it is known as the doomed planet. It is so torn apart by the gravitational tides of its star that it is shaped like an egg and destined to be completely engulfed in its star within the next 10 million years.

By contrast, the coldest world detected so far also happens to be one of the farthest away ever found. It is located 21,000 light-years away from the Earth, toward the center of our galaxy. It was nicknamed Hoth for its resemblance to its Star Wars namesake. The real Hoth is far away from its star, a cool red dwarf, and unlikely to experience any conditions suitable for life. Unlike 55 Cancri e, it is a planet that would not look alien to us. If Hoth were in our solar system, it would be somewhere between Mars and Jupiter, orbiting around its star in about ten years. It is an Earth-like rocky world of about five Earth masses. Its surface temperature of minus 220 degrees Celsius suggests that, if present, volatiles such as water, ammonia, methane, and nitrogen are frozen solid.

Staying with the Star Wars theme, Kepler-16b was nicknamed Tatooine and was the first planet discovered with two suns in its sky. Ten more circumbinary stars were detected since, and while Kepler-16b is unlikely to be habitable, Kepler-453b orbits within the habitable zone of its two-star system. The Cloud City planet Bespin finds similarities in a few other gas giants. Kepler-22b, a super-Earth covered by a mega ocean, would be an excellent analog to Kamino. The molten world of Mustafar finds its doppelgängers in lava-covered Kepler-10b, Kepler-78b, and CoRoT-7b. And the list goes on. Reality is, at least, as strange as fiction, as we now also have to understand what habitability and life mean around the triple and quadruple star systems discovered by Kepler and TESS.

And then there are planets so strange that, theoretically, they should not even exist—which, of course, is the arrogant way of saying how little we really know. In that category of planetary environments, TrES-4 is one of the strangest. About 1.7 times the size of Jupiter, it is one of the biggest exoplanets discovered yet, but has the density of balsa wood. Its mass-to-density ratio is an anomaly that our current models cannot explain and its gravitational pull is so weak that it cannot prevent part of its upper atmosphere from escaping into space, forming a comet-like tail in the planet's wake.

It is often difficult for us to relate to many of these environments because of their extreme departures from the planets we know. Yet, some might still give us a window into our own fate when the solar system will reach its final days. This is the case for WD 1145+017b, a small rocky world about the size of Europa. The planet orbits its parent star in 4.5 hours at a distance of barely 756,000 kilometers, only twice the distance between the Earth and the moon. The star is a white dwarf that ended its main sequence as a red giant. It was probably not too different from our sun before it ran out of fuel and expanded as a planetary nebula now dispersed. The proximity of WD 1145+017b to its parent star provides clues about their possible interactions when the star reached the end of its life as a red giant. The planet's surface temperature nears 3,730 degrees Celsius and rocky minerals are being vaporized off its surface into orbit, forming a disk of hot dust around the star. This progressive vaporization will ultimately lead to the disintegration of WD 1145+017b within the next 100 to 200 million years, a prelude to what is to come for us in a few billion years.

Environmental extremes come aplenty with exoplanets. There are so many of them that we can probably find anything our imagination can conceive. Their oddity is demonstrated by how many are already being compared to fictional worlds. So far, we have discovered planets that reflect less light than coal. On others, it rains molten iron, molten glass, or diamonds. Some planets have clouds of metal vapor or rainfall of liquid rubies and sapphires like on WASP-121 b. Planets such as Gliese 1132 b, like a lizard losing its tail, regrew a second atmosphere after losing its first. There is also a tidally locked world that experiences hellish temperatures on its dayside, and below-freezing temperatures on its nightside. On the other hand, another hot Jupiter tidally locked to its parent star found a way to equalize these temperature differences between day and night sides. It regulates them with insanely fast winds reaching nearly six times the speed of sound, possibly even topping out at 22,000 miles per hour. KELT-9b, a Jupiter-type gas giant, has a surface temperature of 4,300 degrees Celsius. The molecules in its atmosphere are breaking down to the atom level and burning off.

And there is the matriarch of them all, at least for now, the oldest exoplanet discovered so far. PSR B1620-26 b is 12.7 billion years old, formed barely a billion years after the Big Bang. If only planets could talk. And to some extent, they can. Characterizing its composition will help us understand what heavy elements were present at the time of its formation. This critical piece of information may allow us to estimate when the essential building blocks became sufficiently abundant for life to emerge in the universe. By contrast, V830 Tauri is only 2 million years old and still forming. This diversity of worlds reflects a dynamic and wild universe where every possible scenario and stages of planetary evolution seem to play out in front of our eyes all at once. However, our most profound quest lies beyond our awe-stricken imagination discovering worlds we never thought could exist. It all comes down to one question: Is there another Pale Blue Dot hiding amid this dazzling dance of orbs?

SEEKING EARTH 2.0

Our quest for Earth 2.0 is not motivated by a desire to escape. It should never be unless our planet comes under the threat of mass extinction from a cosmic cataclysm. In any case, the worlds we discover are too distant to reach with our current technology, but some bold mission concepts are emerging. One of them is Breakthrough Starshot. The project aims at reaching our nearest planetary neighbor beyond our solar system, Proxima Centauri b, a planet within a system of three, the last one discovered in early 2020 with ESO's VLT. Proxima b has a mass comparable to the Earth and is located in the habitable zone of its parent star. The mission would consist of a fleet of light sail interstellar probes riding on a multi-kilometer array of beam-steerable lasers. Surfing light, it would take them between twenty and thirty years to complete the 4.37 light-year journey. The Parker Solar Probe, our fastest spacecraft to date, flies at 531,000 kilometers per hour (0.05 percent the speed of light) and would take seven thousand years to reach this destination. To put it into perspective, this is more than twice the span of the Egyptian civilization. The technology to make Starshot happen is daunting and requires reaching 15 to 20 percent of the speed of light to succeed. Maybe this is something the next generation will pull off with a breakthrough in propulsion systems.

In the meantime, we continue to tirelessly decode alien planetary environments remotely from the ground and space. What pushes us is this insatiable curiosity that has driven us since the dawn of time, this urge to find out whether we are alone in the universe or not. From what is unfolding right now before our eyes, I will repeat what I so often say. With a minimum of 300 million exoplanets located in the habitable zone of their parent stars in our galaxy alone, thinking that we are alone in this cosmic ocean is simply a statistical absurdity. That being said, we have to figure out what being in the habitable zone means for a planet, including its potential to stay within it and transition from being a habitable world to one that develops life and is capable of sustaining it over time.

For example, our solar system started with three habitable worlds: Earth, Venus, and Mars. Yet, today, we only have evidence for life on one of them. As the sun evolved from its intermediate stage between a protostar and a mid-mass main-sequence star, its size and energy output changed. So did the boundaries of the solar system's habitable zone. The same should be considered for exoplanets and their parent stars, as stellar activity greatly influences the changes that may take place in their habitable zone over time. It then becomes evident that not all candidates are equal.

There are currently a little over sixty confirmed exoplanets in the habitable zone of their parent stars. One is the size of Mars, and the others are between Earth and super-Earth-sized. Some might be affected by solar flares and high radiation levels, reducing their chances to develop or sustain life regardless of their position in their planetary systems. Kepler-438b illustrates this point. A near Earth-sized and rocky exoplanet, it is located almost 473 light-years away from Earth on the inner edge of the habitable zone of a red dwarf and receives the equivalent of 1.4 times the solar flux. It is also blasted by powerful radiation from its star every one hundred days during violent solar storms. These storms are much more powerful than any superflares our sun ever produced, which likely stripped away its atmosphere and sterilized its surface. Without an atmosphere, the planet is exposed to harsh, short UV and X-ray radiation along with charged particles, leaving little hope that anything could survive on it. Here is thus an example of a planet not too dissimilar to ours and located in the habitable zone, but unlikely to be habitable. On the other hand, life may still be able to emerge and survive near a type of star that probably would not have made our priority list until recently: a dying star.

Stars like our sun become white dwarfs after they run out of fuel and have expelled their outer layers. At this stage of their evolution, they are no larger than many planets. Located 117 light-years away from Earth, WD1054-226 is such a star. Dips in light corresponding to sixty-four even-spaced clouds of planetary debris orbit around it every twenty-five

THE SECRET LIFE OF THE UNIVERSE

hours. The best way to explain the regularity of their transit patterns is with the presence of a planet, which, should it be confirmed, would be located in the habitable zone of the white dwarf. Such a star should have cleared its orbit during its main sequence and any planet found in this region of the planetary system has to be recent. Moreover, although the habitable zone of a white dwarf is smaller and closer to the star, a planet in that area can expect sustainable habitability for at least 2 billion years. It took life much less time to emerge on Earth.

TRAPPIST-1

Meanwhile, forty-one light-years away in the Aquarius constellation, the TRAPPIST system allows us to observe the evolution of habitability on several planets located in the habitable zone of a 7.6-billion-year-old ultra-cool red dwarf star barely larger than Jupiter. In this system of seven planets, at least three of them are located in the habitable zone. They range in size from Mars-like to slightly larger than the Earth. They may represent just one example among many such systems in our galaxy, as the Spitzer data suggest a probability of Earth-sized worlds forming around ultra-cool dwarfs to be as high as 30 to 45 percent!

If the TRAPPIST system were put to scale with the solar system, all seven planets would fit within the orbit of Mercury. They are densely packed and in an orbital resonance, meaning that they are spaced with specific numerical ratios and exert periodic gravitational influence on each other. This proximity to their parent star and each other generates tidal interactions that could allow plate tectonics, a critical element when thinking about the emergence of life and its role in biogeological cycles. Due to this proximity, some planets are also visible in each other's sky, but are unlikely to have moons.

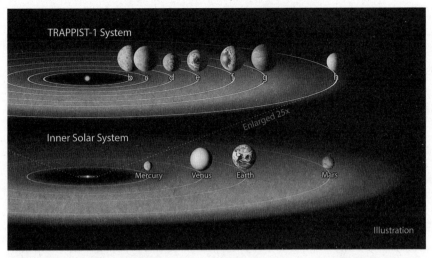

Comparison between the TRAPPIST-1 system and the inner solar system. The TRAPPIST system includes seven planets close to the dimensions of the Earth. Three of them (TRAPPIST-1e, f, and g) are in the habitable zone of their star.

The unique configuration of the TRAPPIST system offers an almost textbook physics and chemistry course in the evolution of water as the distance between planets and parent star increases. At the same time, environmental conditions are complicated by excess radiation from the star's early history. The innermost planets might experience intense volcanism and magma oceans, and the ultraviolet flux from the star probably stripped them of their atmospheres. On the other hand, located farther away, the outer planets may be at a safe enough distance to keep an atmosphere and, for some, interior oceans. The cold temperatures of the farthest planets would prevent surface liquid water in most cases. These worlds could be the equivalents to the snowball Earths, a hypothesis that proposes that 2.2 billion and 635 million years ago our planet was completely covered in ice, although they could be warm enough with a CO_2 atmosphere to reach the melting point. All three most distant planets have the potential for harboring liquid water, with one in particular that could be 50 percent water by mass. The farthest planet from the star in the TRAPPIST system could present two possible environmental scenarios: One would make it a planet very similar to Titan, with a methane/nitrogen atmosphere. The other shows that the melting point of water could be reached if the planet has a hydrogen greenhouse effect or enough internal heat and tidal heating.

The first of the outer planets, called TRAPPIST-1e, is the most Earth-like. It has a similar mass, radius, density, gravity, temperature, and stellar flux as our planet, as well as a rocky surface and possibly a compact atmosphere like the terrestrial planets in our solar system. Moreover, the TRAPPIST-1 star is only 8 percent of our sun's mass, giving it a life span potential of 12 *trillion* years, over one thousand times longer than our sun. These factors appear extremely favorable to the emergence of life, even though the planet is tidally locked to its star. While this may impact its climate and habitability potential, its atmosphere may be strong enough to allow additional heat to be transferred to the night side. TRAPPIST-1e is the closest planet to Earth 2.0 we have found so far, alongside several other exoplanets like Kepler-186f or Kepler-452b, which are the size of Earth or super-Earths detected in the habitable zone of their stars. Life could still hide outside that zone, depending on the type of planets and atmospheres and maybe the presence of moons. The possibilities are endless.

THE ROLE OF ARTIFICIAL INTELLIGENCE IN EXOPLANET DETECTION

Although the data collected with our telescopes may not seem like much for now, knowing the radius, mass, density, and distance to a parent star and the type of orbit enables us to run atmosphere and climate models. With them, we can draw from our knowledge of Earth's evolution and infer what types of environments an exoplanet might experience. We can then compare these environments to the various epochs Earth has traversed throughout its geological and climate history, including the young prebiotic Earth, the great oxidation event, snowball Earth, and more. These represent physicochemical snapshots that help us produce spectral templates to support the search for exoplanets and their characterization. Knowing the range of plausible climates and environments, we can use our understanding of terrestrial organisms, including extremophiles, to evaluate what types of metabolisms and biology could develop, adapt, and survive, and where, and what could be their plausible biosignatures. But we have to do this in a universe that presents us with an endless diversity of habitable zones and environments around stars of many different types. These environments open almost boundless horizons for our search for biosignatures, especially when we are still wrestling with what constitutes an unambiguous fingerprint of life in the environmental background noise.

It is a challenge that artificial intelligence (AI) and machine learning (ML) tools have allowed us to take on much more efficiently in the past decade, fostering headways in exoplanet detection. For instance, AI algorithms are more effective in distinguishing false positives from real signals than humans. They discovered 301 unknown exoplanets in the Kepler and K2 data and validated 50 more. These technologies are also used to model habitable zones, theoretical coevolution of life and alien environments, and interpret data collected by ground-based and space telescopes. They help us tackle evolutionary systems in biology, complexity, and novelty, which are open-ended problems dealing with adaptation and mutations that would be otherwise overwhelming to deal with. AI not only helps us parse through enormous datasets from ground-based and space telescopes, it also guides us in planning future observations and designing new experiments, instruments, and missions.

Detecting and grasping life's diversity from simple to complex cannot, and will not, be achieved with a "one size fits all" approach. Instead, it requires bridging astronomy, biology, planetary and environmental sciences, oceanography, cosmology, the humanities, and much more, and with it, the development of new specialized tools ranging from mega telescopes to nanotechnology. In the nineties, the notion that such bridging was necessary started to emerge, and the NASA Astrobiology Institute and the first astrobiology road map were created in 1998.

Today, technological advances and vast volumes of data provide a higher-resolution, multidisciplinary foundation to map a complex universe populated by more planets than stars. AI brings unprecedented computing power to support detection and modeling. These techniques, emerging in an era of giant telescopes, segmented mirrors, and adaptive optics, will make upcoming ground-based telescopes more powerful than current space telescopes. The Large Synoptic Survey Telescope will detect a wide range of exoplanet populations; the Gemini Planet Imager is imaging them. Soon, the Giant Magellan Telescope will measure their velocities and atmospheres with a resolving power ten times greater than Hubble. Detection comes in CubeSat sizes, too. The Transiting Exoplanet Survey Satellite continues Kepler's hunt; the James Webb Space Telescope is surveying atmospheres in the infrared. With these telescopes, we are better positioned than ever to find life beyond Earth. However, we are still wrestling with what constitutes unequivocal evidence because there is still no consensus definition for what life is.

10

ECHOES OF
COSMIC WAVES

The abundance of exoplanets suggests that living worlds may be already hiding in plain sight in our night sky, which brings hope that the first detection of biosignatures could take place in the coming decade, but for now, the question remains as to whether life, simple or complex, exists beyond our solar system. Thinking of how rapidly simple microorganisms emerged on Earth and how common their building blocks are, one could easily picture a universe teeming with chemoautotrophs and cyanobacteria or their analogs—at least I do. They, too, may leave biosignatures in their planets' atmospheres in the form of biogenic gases and thermodynamic disequilibria. Considering the statistical probability of their existence, they may be the first to be detected. But, somehow we want more. We are sounding the depths of night, hoping that "someone" is looking back, maybe also questioning and searching, maybe already on their way.

Finding a technologically advanced extraterrestrial civilization is not contingent on finding simple life first. Yet, discovering another world populated by microbes would allow us to understand better what conditions lead to complex life on one planet and not on another. Although microorganisms seem to be of interest only to experts, they are foundational to all living beings in our biosphere and to an understanding of alien biology. Too often, we forget that complex organisms are federations of microbiomes living together symbiotically, including us. All life—no matter where it is at in its stages of evolution and complexity, or its ability to interact and adapt to its environment—takes root in these humble beginnings. Essential clues on how to decipher the signatures of alien life reside in decoding the behavior of these diminutive beings. Finding evidence of microbial biospheres beyond Earth would be a critical step forward also because, in this statistical and probabilistic universe of ours, the more we find the likes of them, the greater the odds are that complex, intelligent, and technologically advanced life also emerged beyond Earth.

IN SEARCH OF COMPLEX LIFE: THE DRAKE EQUATION

Since its creation by Frank Drake in 1961, the so-called Drake equation, which is in fact more of a formula, has provided an intellectual framework and probabilistic arguments to think about the conditions needed for life to emerge and to reach the stages of a technologically advanced civilization. While it was meant to narrow down the number of these potentially active and detectable extraterrestrial civilizations in our galaxy, this rationale was already embedded in its various terms.

$$N = R \cdot f_p \cdot n_e \cdot f_l \cdot f_i \cdot f_c \cdot L$$

The Drake equation is a mathematical formula for estimating the number of advanced extraterrestrial civilizations whose electromagnetic emissions are detectable. It was created by Frank Drake on November 1, 1961. It is the most famous equation after Einstein's $E = mc^2$.

In it, (N) represents the number of civilizations in the Milky Way galaxy whose electromagnetic emissions are detectable; (R.) is the rate of formation of stars suitable for the development of intelligent life; (f_p) is the fraction of those stars with planetary systems; (n_e) the number of planets per planetary systems with an environment suitable for life; (f_l) the fraction of suitable planets on which life appears; (f_i) the fraction of life-bearing planets on which intelligent life emerges; (f_c) the fraction of civilizations that develop a technology that emits detectable signs of their existence into space; and (L) the length of time such civilizations emit detectable signals into space.

When the Drake equation was created, all we knew from planets and stars came from ground-based telescopes, and the default solution to (N) was *1,* that is, us. It is still today the only number we can claim as factually true. However, new data overwhelmingly point toward (N) being suitably greater than *1,* The Kepler mission taught us that, on average, every star in the sky has one or more planets orbiting around it, and 10 percent of them are sun-like stars. Most exoplanets discovered so far fall between the size of Mars and Neptune, and approximately 20 percent of them are rocky worlds, Earth-like, and in the habitable zones of their stars. Excluding the mini-Neptunes or any other habitable worlds located outside these categories that could also be habitable, we may be already looking at a few billion Earth-like planets in the habitable zone of their stars in just our galaxy. Sixty years after the creation of the Drake equation, there is thus much greater confidence that, unless our universe is statistical nonsense, we simply cannot be alone, but despite it all, we still have to come up with the first evidence of an extraterrestrial civilization.

The Drake equation was created to evaluate the size of (N). It also proved to be a durable intellectual framework and an instrumental road map to guide us in exploring a universe of potentially living worlds. But there is a lot more to it than meets the eye. Its structure makes it a multi-dimensional glyph of encrypted messages that tells us how many extra-terrestrial civilizations might be out there for us to find. But that's only the beginning, just the most visible part of the formula. It also points to where to search with (f_p) and (n_e); (f_l), (f_i), and (f_c) suggest what stages life must go through from its simple origins to the rise of advanced techno-logy. It also indirectly suggests what extraterrestrials could have evolved into considering the characteristics of the planetary environments they could have emerged from.

In 1961, the formula was conceived keeping in mind Earth-like environments and a type of life that could be familiar to us. Findings from Kepler, TESS, and other spaceborne and ground-based telescopes made this specific quest even more relevant. Yet, the formula is a flexi-ble tool that can absorb the various new notions of habitable zones and exoplanets and concepts of life as we know or might not know it. All exoplanetary environments discovered so far can be passed through its cogs. But like any other intellectual tool, it also has its shortcomings. For instance, the formula shows a structural gap between the number of planets developing intelligent life and those developing civilizations, as if it assumed that all intelligent life produces civilizations. While our planet counts dozens of intelligent species, such as apes and some ma-rine mammals (e.g., dolphins) and many more, and some have developed complex societal organizations, "civilization" usually refers to complex human societies as defined by "the stage of human social and cultural de-velopment and organization considered most advanced." It includes spe-cific criteria such as establishing a stable food supply, a social structure, a system of government, a religious system, a highly developed culture, advances in technology, written language, and the keeping of written records. Regardless of how we feel about this definition, this is what we must work with for now.

In theory, the formula should mention the fraction of planets (f_s) where intelligent species transition from a societal structure to a civilization. This term could also be expressed as a number (n_s) of intelligent species on a planet that transitions from a societal system to a civilization. It is a significant piece of information, since we cannot exclude the possibility that more than one intelligent species could develop a civilization on any given planet. Ultimately, this missing factor could impact our evaluation of how many extraterrestrial civilizations exist and can be detected. But, and equally important, it provides a space to consider that species can be highly intelligent without developing civilizations, as we observe in nature.

It can be then argued that, in light of recent discoveries and scientific advances, the formula could be refined and updated, and many are giving it a try. But, beyond its limitations, it stands as a robust road map and the first universal vision of astrobiology ever created. In that, it will forever remain an enduring testament to that particular moment in time when humanity started to think holistically about the search for life in the universe. Ultimately, it is a brilliant probabilistic argument that, sixty years after its creation, continues to take on new dimensions with the input of new information.

Although uncertainty still increases as we move to the right of the formula, the first three terms to the left can now be populated with more robust data than six decades ago. For instance, ($R \cdot$) is the number of stars formed per year that are suitable for the development of intelligent life. Astronomers now understand that the Milky Way is not the biggest producer of stars in the universe at this stage in its evolution. Only about half a dozen of them are born in it every year. However, while a yearly rate of formation is informative, it is not extremely useful. Instead, it can be advantageously replaced by the total number of stars of interest for the SETI search produced by our galaxy since it was born. The resulting number gives us a better notion of how many such stars the Milky Way produced over time that could have led to technologically advanced civilizations. It represents a theoretical potential.

Let's take sun-like stars as an example. The current estimate of their population in the galaxy is 4.1 to 11 billion, with models suggesting that about 7 percent of them could host at least one habitable planet (but it could be as many as half for them). The population (n_e) of planets in the habitable zone of sun-like stars could then reach 300 to 2,000 million. It is a very large number, but it should not be taken at face value, since not all such planets will maintain a sustainable environment for biology over their lifetime. For example, Earth, Venus, and Mars in the solar system demonstrate how fleeting habitability can be around a sun-like star. The solution must then take into account the evolution of a star and its habitable zone over time, as well as the specific properties of the atmospheres and surfaces of planets orbiting it. Since the end goal of the Drake equation is to evaluate how many detectable civilizations are out there, one fix to simplify the solution for the number of planets in the habitable zone of sun-like stars could be to consider only planets within the habitable zone that demonstrate the potential for long-term sustainable habitability. These would be worlds that give enough time for life to theoretically evolve to high levels of technological advancement. The flaw of this fix is to neglect the possibility that for older stars entering a red giant phase, a civilization might migrate to the deeper ends of its planetary system to seek refuge in outposts located on minor planetary bodies such as moons or asteroids. Importantly, exoplanets also show us that habitability extends beyond Earth-like planets and includes super-Earths and mini-Neptunes, which may substantially increase the number of candidates for long-term habitability.

The most significant remaining uncertainty in the equation lies in our understanding of the ratio between the numbers of worlds where

life emerges and those where it ultimately evolves into advanced civilizations. Compounding the issue, many questions remain unresolved today
surrounding life's origin. We do not know what life is. We still debate
how it started on our planet, and we have yet to figure out whether it
appeared elsewhere in our solar system. A temporary solution to the fraction of planets on which life has actually appeared (f_l) is to use the only
evidence we have. In our solar system, life appeared on the one planet that
preserved sustainable habitable conditions over billions of years of environmental changes within the habitable zone of a sun-like star: Earth.
Mars did not maintain habitable conditions for very long, at least on its
surface. For Venus, the jury is still out on how long habitability endured.
For both Mars and Venus, the existence of past and present life has not
been ruled out yet. So, the fraction of planets where life emerged in the
inner solar system could range from one to three planets out of three in a
changing habitable zone. Further, we cannot exclude the possibility that
it also emerged in the outer solar system's habitable environments.

In trying to approximate the fraction of life-bearing planets where
intelligent life emerges, we also have yet to agree on the meaning of
intelligence, and still struggle with many versions of its definition. Its
broader designation is "the ability to acquire and apply knowledge and
skills." In nature, adaptation could loosely fit this description. In this
case, microbial species that survived and adapted to numerous environmental changes over billions of years could be deemed intelligent. However, although microbial species may release biosignatures detectable
from space, it is not through the development of technology, but as a
consequence of natural metabolic activity.

A more comprehensive definition is "the ability to learn, or understand, or deal with new or trying situations, the skilled use of reason,
or the ability to apply knowledge to manipulate the environment, or to
think abstractly as measured by objective criteria." This definition from

Encyclopedia Britannica brings us closer to what can be applied to the complex behavior of species such as primates, some marine mammals, octopi, rats, crows, and more. It also provides a pathway to the following term in the equation (f_c), since some intelligent species have developed societal structures. One of them, humans, ultimately developed diverse civilizations over time. The most recent has developed technologies capable of emitting detectable signs of its existence into space.

The last factor in the equation (L) probably bears the most uncertainty of all the terms. It evaluates the length of time a technologically advanced civilization broadcasts detectable signals into space. If we follow our evolution on Earth as a representative blueprint, life emerged early on our planet but it remained very simple for billions of years. Humans, as a species, are very latecomers, and modern humans have been walking the Earth for only 0.0017 percent of the planet's history. A common metaphor compares the entirety of Earth's history since its formation to a twenty-four-hour clock. On it, modern humans made an entrance on the planetary stage just a second before midnight. And, although modern humans appeared 75,000 years ago, we, as a species, entered the industrial era only in the past 250 years. Then our civilization began to unwittingly and carelessly alter the signature of the Earth's atmosphere at a detectable scale. Aviation was born 120 years ago; Sputnik 1 launched in 1957; humans landed on the moon in 1969 and are now thinking of building outposts on the Moon and Mars. Technological advances and progress have been exponential in this very short time window. However, unless evolutionary biology operates very differently on other planets, it takes billions of years for life to transition from simple forms to technologically advanced beings. If we accept this hypothesis as a universal rule for habitable planets around sun-like stars and assume that evolution follows a similar trend elsewhere in the galaxy, some conclusions become self-evident.

POTENTIAL CANDIDATES FOR TECHNOLOGICALLY ADVANCED CIVILIZATIONS

The dawn of technologically advanced civilizations broadcasting detectable signals and being able or willing to communicate takes place around middle-aged sun-like stars. Within 5.5 billion years of the star's lifetime, those civilizations will have to relocate when their star's expansion transforms their planetary atmosphere and the environmental conditions are not survivable anymore. This is the fate awaiting our planet 1 billion years from now, when a lack of oxygen will wipe life out. It will happen as the sun becomes more luminous and more energy reaches the surface of the Earth. The process will accelerate the weathering of silica rocks, altering all biogeochemical cycles and the interactions between the mantle and surface environments. It might not be the end of all life on Earth, but that time will undoubtedly mark the end of all complex, aerobic life-forms that cannot escape.

Sun-like stars younger than our sun could, in theory, become suitable to the development of future civilizations. Yet, their young age makes them unlikely to harbor a detectable civilization unless one of them migrated to such a young system to seek refuge or expand. Considering those stars that are too old or too young, the window of time to search for advanced civilizations around such systems is probably when sun-like stars are between 4.5 and 5.5 billion years old. Obviously, this is an entirely Earth-centric model that requires converging evolution, and it may or may not reflect reality. It still makes sense considering how common the building blocks of life are and how long it took for simple life to evolve to complex forms on our planet. And, if not, the sheer number of sun-like stars makes it statistically almost inevitable that this scenario played out in our galaxy and beyond more than once, even just by accident.

Coincidentally, sun-like stars started to form 10 billion years ago, and their life span is 10 billion years. The oldest technologically advanced extraterrestrial civilizations around such stars could have thus arisen when our planet formed and life emerged on it. Many probably have already disappeared, while most of those left are likely much more advanced than we are. Considering our scientific and technological progress compared to two hundred years ago, one can only imagine at what stage of development an alien civilization just five hundred to one thousand years ahead of us could be.

Conversely, only a fraction of that which developed in parallel to us may be already detectable today. Some of these newcomers could be located somewhere around stellar siblings to the sun formed in the same gas cloud and the same stellar cluster 4.6 billion years ago. One of them was actually discovered recently. HD 162826 is located in the constellation Hercules, about 110 light-years away from Earth. It is just 15 percent more massive than our sun and has 3 percent more metals in its composition. Although none have been detected yet, terrestrial planets are likely to have formed around it, considering the star's metallicity.

HD 162826 could be an intriguing candidate for SETI researchers. The star is a sibling of our sun, very similar in composition, and born at the same time 4.6 billion years ago. Its distance from Earth brings it within reach of Earth's 230-light-year-diameter radio bubble, the maximum distance reached by our first radio emissions. In a perfectly analogous (and highly anthropocentric) scenario where a technologically advanced civilization arose around HD 162826 at about the same time as we did and is now looking and listening, the first contact we might ever make could be with our own stellar family! It also demonstrates how recent technological civilizations like ours may be relatively isolated, the echo of our radio waves barely starting to form ripples in the fabric of the cosmic ocean. On the other hand, this sense of isolation may simply result from a human-centric reflection. Older civilizations are likely to have more advanced knowledge of the laws ruling the universe. They could already be eavesdropping on our communications without our knowledge. And there might be many of them.

Current estimates of the number of planets with advanced civilizations include the modeling of data from exoplanet detection, extrapolations on physicochemical environments, evolutionary biology, and societal evolution on Earth. Intellectual license is taken to use our solar system's evolution, terrestrial biology, and human evolution as a blueprint for what may be happening elsewhere in the galaxy. They do not imply, though, that the extraterrestrials we might find will be human-like. Instead, reviews of Earth's history in the context of the Fermi paradox[1] simply assume that, if given time, intelligence will emerge on a fraction of the planets where life started and will ultimately lead to technologically advanced civilizations releasing detectable signals. How often this may happen and how many civilizations will appear is still open to substantial subjectivity. The only example we have is ours, one data point.

Adding to the complexity of this evaluation is the role of random events such as natural or induced catastrophes, the lingering unresolved questions about what altered the course of biological evolution many times, and other factors that still remain challenging to quantify. Further, a study by Luke Kemp in 2019 shows an *average* life span of 340 years for ancient human civilizations, which represents a very short window in time. Some of them lasted much longer (for instance, the ancient Egyptian, Vedic, Olmec, Phoenician, Carthaginian, Kushite, and Minoan civilizations). Others were very short-lived (for example, the Akkadian and Phrygian empires). How accurate this life span is for a modern human or an alien technological civilization is unknown, but this study (and others[2]) could suggest that collapse may be a normal phenomenon for civilizations, regardless of their size and technological stage. On the one hand, technological and scientific advancements have cursed our species with the ability to eradicate itself at any given time now. On the other hand, they foster remarkable progress that may prolong individual life spans. At the scale of our civilization, they bring us closer each day to the moment we will spread into the solar system, increasing the chances we may survive away from our original planetary cradle.

Determining how many advanced civilizations may be beyond Earth is further complicated, as technologically advanced civilizations may not be detectable for various reasons. Assuming they all remained detectable and all factors are accounted for, including the sizable grain of salt such estimates should be taken with considering our current state of knowledge, the number of technologically advanced civilizations around sun-like stars could range between a few dozen and a few thousand depending on models. This estimate can probably be increased manyfold, since it only considers Earth-like planets located in the habitable zone of sun-like stars.

Other types of worlds presenting favorable environments despite not being Earth-like have broadened our perception of habitability and expanded our parameters for detection. For instance, we are barely starting to envision what climate and habitability would mean on a planet orbiting a system of two, three, or four stars. Could, or did, life emerge on any of those? Maybe. Gravitational interactions within a multi-star system may lead a planet to be ultimately flung out of orbit, crash into its star, or may decrease the extent of the habitable zone because of orbit instability. Yet, mathematical models show that multiple systems like Kepler-34, -35, -38, -64, and -413, including circumbinary giant planets, have permanent and stable habitable zones. They could support life as we know it, and some may even have an ocean. Seasons on these worlds would likely be considerably more variable than on Earth on a timescale of thousands to tens of thousands of years, including periods when seasons may disappear altogether because multiple gravitational tugs trigger variations in tilt of the planet as it orbits. This setting would generate a different style of coevolution of life and environment, and the consequences on metabolism and the development of complex life still have to be understood. But maybe, just maybe, someone is looking at multiple suns setting on the horizon of a distant planet right now and wondering if someone is looking back, too.

Another reason for the number of planets with advanced civilizations to substantially increase is the existence of other stars, like M dwarfs and K-type stars, which may provide a suitable environment for life, too. With only a fraction of the sun's mass and luminosity, M dwarfs are among the smallest and coolest stars. Their life spans 2 *trillion* years, but their love-hate relationship with habitability makes it a challenge to evaluate their potential to develop life. They are much more variable than our sun, with giant stellar spots and massive flares triggering dramatic dimming and brightening. Their unpredictable X-ray and ultraviolet environment may annihilate life's chances to emerge early on planets orbiting them. Nonetheless, they are so abundant—there are as many as possibly 58 billion of them—that the probability that at least a subset of them could harbor planets with life is real. In this case, time could be theoretically on the side of complex evolution.

K-type stars (orange dwarfs) represent other high-priority candidates for fainter and cooler stars. Although their habitable zones may be small, they could plausibly host life-bearing planets with lifetimes ranging between 15 and 45 billion years, compared to 10 billion years for our sun. Their radiation environment is more favorable than with M dwarfs, and they may represent as much as 13 percent of the stellar population in our galaxy. With their life span, habitability and life could be beyond our wildest imagination.

THE FERMI PARADOX

We can see how almost three decades of exoplanet investigation have painted a vision of a galaxy full of potential for living worlds. We are just beginning a wondrous cosmic odyssey with NASA's JWST, ESA's PLATO (PLAnetary Transits and Oscillations of stars) space telescopes, and the next generation of cutting-edge ground-based instruments. In striking contrast, this exoplanetary wealth appears to make the question raised by the Fermi paradox—"Where is everybody?"—even more relevant today. The potential is so high for advanced alien civilizations and for them to be older than we are, so why don't we see any sign of their presence in the universe?

The physicist and Nobel laureate Enrico Fermi raised this question in 1950 during a conversation with some of his colleagues. The rocket scientist Konstantin Tsiolkovsky had similar interrogations almost two decades earlier while thinking about space travel. He came up with the "zoo hypothesis" as an explanation. In it, humanity was deemed too young and not ready to be contacted by more advanced beings. Instead, extraterrestrials let evolution naturally unfold without intervening, just like people observing animals in a zoo. Leaving aside for a moment the claims that the signs of extraterrestrial visits exist, let's examine the arguments and possible explanations for the Fermi paradox.

The abundance of exoplanets discovered in our galaxy and the estimated fraction that could lead to the emergence of life seem to support the "mediocrity principle," which states that when a subset of a statistical population is abundant, the probability of randomly sampling one of its representatives is high. In other words, according to this principle, Earth and its evolution represent a common occurrence in the universe, which justifies an anthropocentric approach to the search for advanced extraterrestrial civilizations.

Let's now consider when the first stars suitable for the development of intelligence formed in the Milky Way. The majority of advanced alien civilizations should be older than us, as shown by the example of sun-like stars. Thus, extraterrestrials should have colonized their planetary systems by now, spread to new habitats to harvest new resources, and explored and colonized beyond their close neighborhood. They possibly could have hopped from one planetary system to the next as generations passed. In other words, despite the distance, extraterrestrials should have been knocking on our planetary door a long time ago. Instead, we are facing the "Great Silence."

A couple of decades after Fermi's death, in 1954, the astrophysicist Michael Hart suggested possible solutions to the paradox. One links the lack of evidence with obstacles to space travel due to biological or technological limitations. Another considers that despite the age of the universe and suitable early stars, advanced civilizations may not be much older than we are and do not have the capabilities to reach us yet. It is also possible that extraterrestrials visited the Earth, but we do not know about it. Finally, extraterrestrials might simply not care if life is out there at all.

THE RARE EARTH HYPOTHESIS

Many other studies have suggested solutions to the Fermi paradox since and the types of events that could act as a "Great Filter," precluding us from making contact. In the context of the Fermi paradox, it is whatever prevents non-living matter from becoming living matter. Thinking that intelligent life is abundant in the universe assumes that the universe is similar in all directions at the macroscopic scale, consistent with the Copernican principle and the mediocrity principle, but the paleontologist

Peter Ward and the astronomer Donald E. Brownlee thought there could be more to the Great Filter when they proposed the Rare Earth hypothesis.[3] They agreed that simple life should be common in the universe. However, they concluded that advanced life is rare because of a series of filters that create obstacles preventing complex life from emerging, making Earth a unique place.

The Rare Earth hypothesis was proposed a few years before new telescopes and spaceborne missions revolutionized our vision of the abundance and diversity of exoplanets in our galaxy. Some factors clearly do not have as much weight today as Ward and Brownlee thought they would when they proposed them. For instance, we know from the Kepler mission that the fraction of stars with planets in our galaxy is considerable (superior to 0.8) and the fraction of rocky planets staggering. Among those, many are located in the habitable zone of their stars. Planetary systems with large gas giants are the rule, not the exception. Further, like in the Drake equation, other factors are still left to best guesses at this point. Those include the fraction of planets where microbial life and complex life arise. In addition, the fraction of a planet's life span where complex life is present is somewhat ambiguous, as complexity can be relative.

An intriguing aspect of Ward and Brownlee's series of filters is the fraction of stars in the galactic habitable zone. While the notion of a habitable zone is commonly used for planetary systems, its role at a galactic level is more rarely mentioned. The Milky Way might not be the largest galaxy, but it is still a vast place, spanning 180,000 light-years. In it, the solar system is tucked away in Orion's arm, a reasonably quiet region 27,000 light-years away from the center. But there are also some very bad neighborhoods in our island in the sky. The galaxy's center is one of them, blasted by radiation levels unlikely to enable life. Being too close to the center of the galaxy introduces a number of hazards. Our solar system is surrounded by the Oort cloud, a region full of comets. Episodically, these objects are being flung out of orbit by the gravitational tug of a passing star, and they start their journey toward the inner part of the solar system, where they sometimes wreak havoc by colliding with planets. If the Earth was closer to the center of the galaxy, this would happen more often due to a higher density of stars, triggering more collisions and potentially more extinction-level impact events.

Further, the spatial distribution of stars in the Milky Way generally coincides with their populations. The oldest stars (Population III) are found in the Milky Way halo. With no metals, they could not have led to life as we know it and were unlikely to produce Earth-like planets. Population II stars (with few metals) are in the halo and the bulge, some at the periphery. Population I, the youngest stars with the highest metallicity and the population our sun belongs to, are mainly located in the spiral arms. It could actually be used as a counterargument to Ward and Brownlee in that the youngest stars, those with the highest level of metallicity, thus most likely to form rocky planets and develop life, are also those that formed in the quiet region of the galactic habitable zone at a time when the building blocks of life became more abundant. So maybe there is no accident after all. Here we are looking again at the idea of a possible generational aspect to life—at least as we know it. That is already plenty to work with, and it suggests that Earth-like planets with elemental building blocks of life similar to ours might not be that rare at all.

They also suggest that planets with large moons must be rare, and the gravity of a large satellite is necessary to stabilize the planet's tilt, which otherwise would be chaotic, probably making the evolution of complex life on land impossible. We know the environmental impact of the moon on our planet and its role on metabolism and physiology for life. It certainly imparted a rhythm to biological processes once life had emerged. However, a counterargument to the suggested rarity of planets with large moons came recently from the modeling of giant impactors such as those that may have formed the moon, showing that their formation in similar circumstances may actually be common in other planetary systems. Our moon certainly made Earth more habitable by moderating our planet's wobbles and stabilizing its climate. Thus it had a role to play in who we are today, but how much is still an open question.

The Rare Earth hypothesis also makes a case for the importance of two previously mentioned climate events in the history of the Earth known as the "snowball Earth," when the Earth's oceans and landmasses were covered in ice from the poles to the equator. Each of these periods lasted for about 10 million years. The first one took place nearly 2.2 billion years ago and the second 635 million years ago. Both episodes coincided with watershed moments in life's evolution. The first was the Great Oxygenation Event; the other was the Cambrian explosion. The first global glaciation was synchronous with the rise of photosynthetic life, which ultimately injected oxygen into the atmosphere and reduced greenhouse gas levels. The second one coincided with the emergence of most animal lineages. These events were what directed Earth's biological evolution on the path that led to today's biosphere. But there, too, it might be more complex than it seems. While the Cambrian explosion remains largely a mystery, and looks like a singularity in the evolutionary picture, it could just as well have emerged as the result of small, discrete environmental changes over long periods of time that ultimately led to major evolutionary thresholds. Recently published work demonstrates that complex life (eukaryotic protists) already existed 2.1 billion years ago, which changes the perspective on the speed at which life transitioned from simple to complex.

Additional counterarguments were raised since the publication of the Rare Earth hypothesis and its recent reformulation by the French astronomer J. P. Bibring, including that some exoplanetary environments may be more favorable to life than the Earth. For example, so-called super-habitable planets may orbit around K-type stars, their life span extending between 18 and 34 billion years. Plate tectonics may start on those planets that reach two Earth mass and 1.3 Earth radii. If they are nestled in the habitable zone of their stars, they could experience livable conditions for a very, very long time. These characteristics have implications for the density of their atmospheres, their topography, and the size of their oceans, which would lead to more shallow marine biodiversity than the Earth.

Ultimately, while the Rare Earth hypothesis in its original or current version offers an interesting intellectual framework to think about, its main issue is that its factors more or less describe an exact replica of the Earth, which turns the hypothesis into a circular argument. Since it is based on a conditional probabilistic formula, no one should be surprised, then, that the result is a very small number and requires a carbon copy of our planet. Following this principle, there could be trillions of galaxies in the universe, and the probability of finding an exact twin of the Earth and of humanity is, indeed, extremely small by virtue of all the random events that led to the evolution of our planetary system, our planet, and our biosphere. But, in its expression, the Rare Earth hypothesis does not leave any room to consider alternative paths toward life and complexity. Its underlying assumption is that the Earth is the optimal model for a coevolution of life and environment—and maybe the only model at all. It does not consider the possibility of alternative biochemistries or similar biochemistries with distinct characteristics, and it is much too early to refute these notions. In this phase of our journey, we have to remain humble and remember that we are barely getting started on our quest for life beyond Earth. And here, the Rare Earth hypothesis can be likened to stumbling upon a beautiful mansion with many rooms, peeking inside the vestibule from the front door, and, upon not immediately finding anyone inside, deciding that there is nobody home.

RADICAL HYPOTHESES

Over time, hypothetical explanations of the Fermi paradox have also included the rarity of extraterrestrial life and extraterrestrial intelligence, and periodic extinctions caused by natural events. In addition, evolutionary explanations propose that intelligent alien species may not have yet developed advanced technologies, which somewhat echoes the idea of a generational emergence of life in the galaxy. In that case, life could just be reaching technologically advanced stages now. Comparing the state of our technology to what it was two hundred years ago, civilizations barely younger than ours might not be in a position to be detected or to communicate just yet.

The aggressive nature of the human species could also suggest that it might be in the nature of technologically advanced civilizations to destroy themselves and possibly export war to other planetary systems. In other words, advanced civilizations are not abundant because they destroy themselves and destroy their neighbors. A related scenario is envisioned in the Dark Forest theory, which is inspired by Cixin Liu's science fiction novel. According to the theory, all life has a strong instinct for self-preservation, and since it is impossible to know whether other life-forms can or will destroy us if given a chance, other civilizations must be treated as potentially hostile. In doubt, the safest option for any species is to annihilate other life before they have an opportunity to do the same. Then we can imagine alien civilizations lurking stealthily in the depth of space.

The Fermi paradox also factors in the brief window in time during which radio emissions could be detectable due to the limited life span of a civilization. Other explanations include socioeconomic behaviors, such as a lack of interest in colonization, lack of resources, societies developed in starships and space habitats rather than on planets, and the fact that it is easier and cheaper to send exploratory robotic probes and machines than biological entities. Critically, we should also remember that we are only at the beginning of our search for extraterrestrials and we are technologically limited. For example, our current approach requires us to be synchronous in space and time to intercept a message. This may change with new search programs.

Beyond our technological limitations, aliens may be so different from us that it makes it impossible to recognize a message that we could have already intercepted without knowing it, simply because they could process information differently than we do. Carl Sagan developed this argument and, if true, it could make us unable to distinguish an alien message from the cosmic background. And the speed at which we process data is only one aspect of the question. How we organize and manipulate them, which does relate not only to our technology but also to the structure of our brains, may make alien communication hidden to us, an explanation proposed by the physicist Paul Davies in 2010.[4] The development of data-parsing methods using AI tools may considerably improve our ability to detect patterns that human brains could have difficulty identifying.

Variations on this idea have multiplied since. Paraphrasing Arthur C. Clarke, who said that "sufficiently advanced technology would be indistinguishable from magic," futurists who believe in machine singularity predict that alien life could be so beyond our comprehension that we could be staring at it without even recognizing it.[5] Machine singularity is a hypothetical point when technology and machine growth become uncontrollable and irreversible, resulting in unforeseeable changes to human civilization. One example would be analog life created by machines that are capable of decoding its blueprint and encoding it on different substrates than its original building blocks. In one extreme scenario, it would be impossible to distinguish artificially produced original (natural) life; in another, the behavior of standard physics could be caused by something else altogether: for instance, the behavior of cosmic matter attributed to dark matter could be a living state instead, which manipulates luminous matter for its own purpose. We would not have any way to tell alien life from the laws of physics with our current state of knowledge.

Before we roll our eyes at the prospect, think about initiatives that are in development, including Elon Musk's Neuralink brain chip that could be implanted in human brains to enhance abilities, or the encoding, transfer, and storage of digital data already performed by DNA. We became a digital civilization only a few decades ago. Imagine what another hundred or thousand years could bring. Futurist scenarios do not come out of the ether. They simply project on alien civilizations one of many potential evolutionary trajectories for humanity itself. Pushed to the extreme, machine singularity becomes the sci-fi-inspired "Berserker Hypothesis," where a civilization creates machines that destroy it and any other life it encounters as it migrates through space. Several versions of this idea have stemmed from the academic world and fantasy novels and provide an alternative explanation to the Fermi paradox by creating another Great Filter.

Yet, despite it all, extraterrestrials are likely to be out there, many versions of them. Whether they know about us or not, or they just ignore us, it may just be that "the absence of evidence is not the evidence of absence," as Sagan said, but simply the reflection of how young our search is.

11

CONNECTING
BLUE DOTS

Let's face it: we are all extraterrestrials. Whether us, born here on Earth, or others we have yet to meet on faraway cosmic dwellings, the building blocks that make us and all the planets were manufactured eons ago in distant stars now long dead. That alone gives us membership in a potentially large galactic family, one with which we long to connect.

SCENARIOS FOR ADVANCED EXTRATERRESTRIAL CIVILIZATIONS

Although we have not met extraterrestrials in person yet, we have imagined them countless times in thought experiments as a way to develop detection strategies for SETI. The reason is that when it comes to finding extraterrestrials, planetary and stellar environments are just the beginning. The Fermi paradox and other theories inventory the plausible causes of the Great Silence. In parallel, we think about what extraterrestrials could do as a way to infer where to search for them and how. This step is essential, as some civilizations may have reached a development stage where they control the natural environment in such a way that it is probably altered beyond recognition.

One of the earliest and most famous studies of this kind was conducted by the astrophysicist and radio astronomer Nikolai Kardashev (1932–2019), who classified possible stages of civilization by how they might harness energy at levels far beyond any human capabilities. Kardashev characterized these stages on a three-level scale that bears his name. On it, humans have yet to reach Type I, a planetary civilization capable of harnessing and storing the energy available on its planet on a global scale. Type II would be a stellar civilization using and controlling the energy of its planetary system, and Type III is a civilization harnessing the energy of an entire galaxy. The Kardashev scale relates to SETI by providing a proxy to determine the odds of detecting these civilizations through the volume and type of information they could broadcast and transmit. Kardashev ultimately concluded that the odds of discovering a Type I civilization were low compared to detecting those who could transmit within an entire galaxy or perform intergalactic communications like Type II and III civilizations.

The Kardashev scale also gives us an indirect approach to thinking of the technologies necessary to control and exploit such energy levels. These could represent the types of technosignatures SETI could be searching for, or, in other words, the measurable properties or effects providing scientific evidence of past or present alien technology. In this case, the presence of a Type I civilization could be betrayed by detecting infrastructures for renewable energy or by constellations of artificial rings of space-based communication satellites. Also known as Clarke Belts, these hypothetical structures are named after the science fiction writer Arthur C. Clarke. A Type II civilization could build a Dyson sphere, a megastructure also made of satellite constellations but, this time, created to enclose a star and extract its energy.

Overview effect. Our generation is the first to see the Earth from space, the first to see its fragile ecosystem in which everything depends on each other. This magnificent view of our planet was taken by an astronaut aboard the International Space Station as it passed over Yemen. Clouds cast long, dramatic shadows.

Terrestrial biosignatures. Each summer, phytoplankton spreads into the North Atlantic and Arctic in blooms spanning thousands of kilometers, like here in the Gulf of Finland. Ocean vortices may bring nutrients from the depths. These types of space images represent one of the many signatures of life on Earth.

10/03/2015 18:42:26 - 85.7 Km

Comet 67P/Churyumov-Gerasimenko
NIR, Orange and Blue Filter from OSIRIS NAC Camera
Credits ESA/Rosetta/MPS for OSIRIS Team - Processing by Giuseppe Conzo

Comet 67P/Churyumov-Gerasimenko. This image of comet 67P was captured by the Rosetta mission on March 10, 2015, from a distance of 85.7 kilometers. Comets are rich in volatile elements and organic molecules that could have contributed to the emergence of life on Earth from impact processes.

Venus. The surface of Venus seen from the Venera 9 probe on October 22, 1975. The color is reconstructed using the Venera 13 palette.

Sunrise over the Meridiani Planum on Mars. Image captured by the Opportunity rover on May 6, 2004, 101 days after the start of its mission.

Martian delta landscape. The delta in the Jezero crater explored by the Perseverance rover reveals a dynamic history of wet periods of deltas, lakes, and rivers, as well as dry periods. Sedimentation, erosion, and volcanism shaped the terrain, which is being further eroded by wind today.

The slopes of Mount Sharp. After over a decade at the Gale crater on Mars, Curiosity is now exploring Mount Sharp's stunning landscapes, like this pass among ancient, eroded deposits.

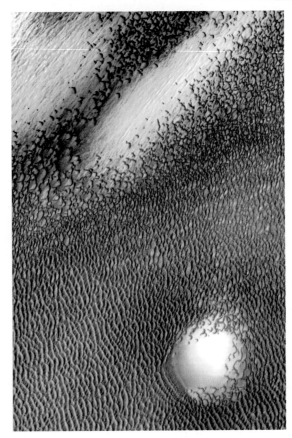

An ocean of dunes. These wind-sculpted dunes surround Mars's northern polar cap, spanning an area slightly larger than France. This image captures a thirty-kilometer-wide region located at 80.3N/172.1E. Blue areas indicate cold, while yellow and orange are warmer.

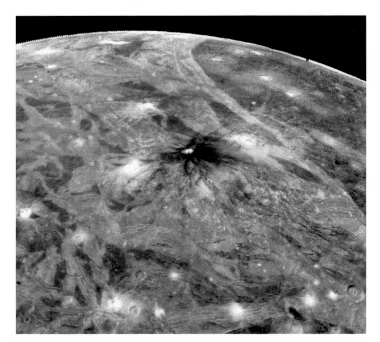

Ganymede. In June 2021, the Juno spacecraft captured an image of the fifteen-kilometer Kittu crater on Ganymede. The crater features a central peak and a bright rim, with faint dark lines indicating the remnants from a past asteroid or comet impact.

Europa. A recent close-up of Jupiter's moon Europa, captured by Juno on September 29, 2022, reveals a young surface marked by ridges and faults shaped by Jupiter's gravitational tides. Their morphology is an additional clue suggesting the presence of an inner ocean.

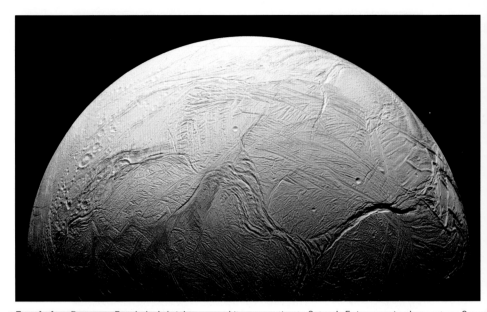

Enceladus. For years, Enceladus's brightness and its connection to Saturn's E ring remained a mystery. Cassini's findings revealed that both find their origin in Enceladus's geyser activity, which is linked to the presence of an internal ocean. Further findings argue that hydrothermal activity in this ocean might support prebiotic chemistry, and possibly life.

Polar Vortex on Titan. This image captured by the Cassini spacecraft shows Titan's south polar vortex and a huge cloud of ice in the orange atmosphere of Saturn's largest moon.

Titan's "Magic Island." Cassini radar images reveal a changing feature in Ligeia Mare on Titan, referred to as the "Magic Island," which seems to periodically appear and disappear, likely due to periodic coastal waves. Similar activity was observed at Kraken Mare. These findings mark the first evidence of dynamic processes in Titan's lakes and seas.

The Occator crater. This colorized image of Ceres's Occator crater (ninety-two kilometers in diameter) features Cerealia Facula, a fifteen-kilometer-wide bright area rich in salts. Cerealia Tholus is visible in its center. The dome is 3 kilometers wide and 340 meters high and nested within a 900-meter depression.

Pluto's atmosphere. Pluto's nitrogen atmosphere, although only 1/100,000th thinner than Earth's, is intricate, as demonstrated by its stratified layers captured by the New Horizons spacecraft on July 14, 2015. The image also shows Sputnik Planitia, a plain of soft nitrogen ice, at the edge of the mountain ranges.

The frozen canyons of Pluto's north pole. Pluto's polar zone features long, wide canyons, the largest being seventy-five kilometers wide, with a winding valley on the canyon floor. Irregular pits, up to seventy kilometers wide and four kilometers deep, suggest melting or sublimation of underground ice. The enhanced color shows a unique yellow hue, unusual on Pluto, due to abundant methane ice and relatively low nitrogen ice, as revealed by New Horizon's infrared measurements.

James Webb Space Telescope's first deep-field image. The telescope captured SMACS 0723, a cluster of bright galaxies, each containing numerous stars, black holes, and most likely planets. Their collective mass acts as a gravitational lens, distorting background galaxies located farther away and allowing us to observe them.

At the heart of the Ghost Galaxy. JWST's image reveals the core of M74, the Ghost Galaxy, showcasing its majestic arms and delicate gas filaments extending from its center. Using mid-infrared observations, JWST aims to study the galaxy's early star formation processes.

Cosmic cliffs. The near- and mid-infrared instruments of JWST have revealed new star-forming areas within the Carina Nebula. What resembles moonlit craggy mountains is actually the edge of NGC 3324, a region of young stars known as the "Cosmic Cliff." This massive gas structure spans 7,600 light-years.

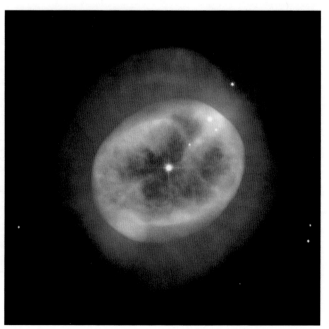

Planetary nebula NGC 2022.
NGC 2022 is a planetary nebula in the Orion constellation. An aging star in its center expels a vast gas envelope into space. As the star's core contracts and heats up, it emits ultraviolet light, causing the surrounding gases to glow.

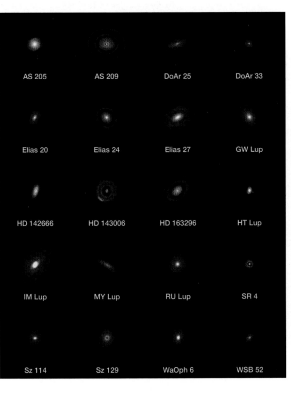

AS 205   AS 209   DoAr 25   DoAr 33

Elias 20   Elias 24   Elias 27   GW Lup

HD 142666   HD 143006   HD 163296   HT Lup

IM Lup   MY Lup   RU Lup   SR 4

Sz 114   Sz 129   WaOph 6   WSB 52

Protoplanetary disks captured by ALMA.
The Atacama Large Millimeter/ submillimeter Array (ALMA) in Chile observed protoplanetary disks in 2018, which will soon become targets of detailed analysis by JWST. These disks represent planetary systems in the making.

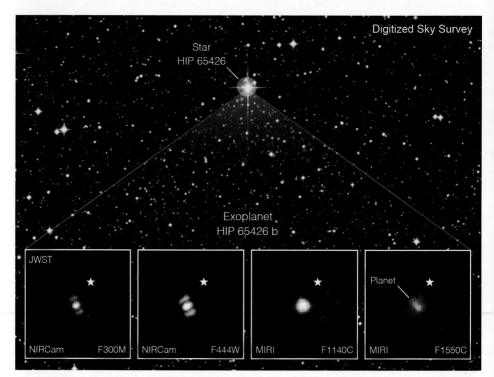

HIP 65426b. This image from JWST shows exoplanet HIP 65426b in various infrared light bands. To reveal the planet, the host star's light was blocked with coronagraphs. The small white star in each image indicates the star's location. The bar shapes in NIRCam images are artifacts, not actual objects.

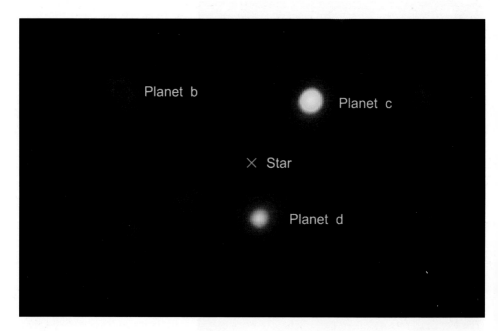

HR8799. Direct imaging around the star HR8799 using the Vortex Coronagraph on the Hale Telescope.

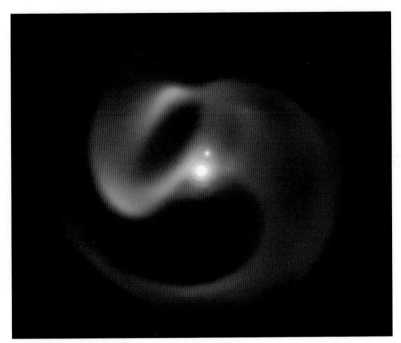

In the Apep star system, cosmic dust forms when powerful solar winds from massive stars in binary or multiple Wolf-Rayet systems collide. Their spinning motion creates a distinctive windmill pattern. Understanding the origin of the dust will deepen our understanding of the development of stars, planets, and life.

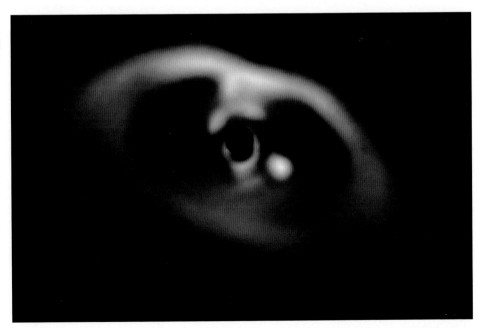

The birth of a planet. This image, captured by SPHERE on ESO's VLT in Chile, reveals the birth of a planet around the star PDS 70. The planet is the bright spot on the right, while the light of the star is blocked by the dark central disk.

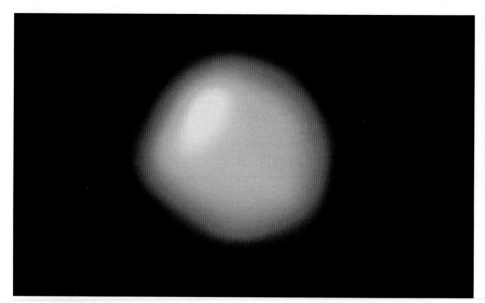

Betelgeuse is a massive, unstable star located in the Orion constellation. It is 1,400 times larger than our sun and its irregular mass loss and brightness fluctuations suggest that it is nearing the end of its life and may explode as a supernova. This high-resolution image shows variations in its envelope and temperature. Studying the elements produced by such explosions and their recycling informs our understanding of the origins of pre-biotic chemistry and life.

Comet2I/Borisov, discovered in 2019, became the second interstellar visitor after 'Oumuamua. Its study by the Hubble telescope provided valuable insights into the chemical composition and characteristics of planetary system ices beyond our own.

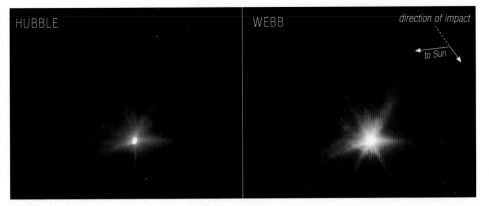

HUBBLE

WEBB

direction of impact

to Sun

Hubble and JWST simultaneously observed the aftermath of DART's impact on Dimorphos, marking their first joint observation. The images capture the Didymos-Dimorphos system shortly after the probe's impact. This mission tested the strategy of redirecting potentially Earth-threatening asteroids to prevent collisions, with DART successfully altering Dimorphos's orbital period by two minutes, confirming the strategy's effectiveness.

The Black Marble. In this nighttime view of Earth known as the "Black Marble," Africa, Europe, and the Middle East are illuminated. The city lights represent one of the most obvious technosignatures that might be visible to extraterrestrial visitors.

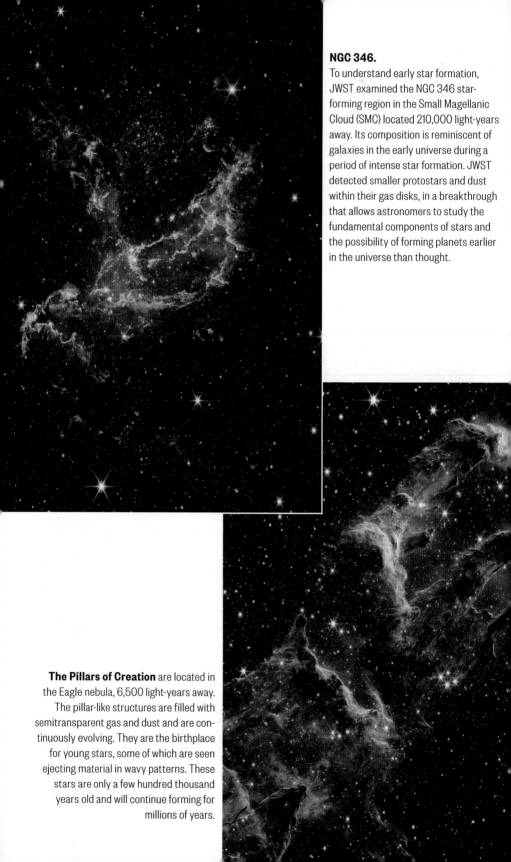

NGC 346.
To understand early star formation, JWST examined the NGC 346 star-forming region in the Small Magellanic Cloud (SMC) located 210,000 light-years away. Its composition is reminiscent of galaxies in the early universe during a period of intense star formation. JWST detected smaller protostars and dust within their gas disks, in a breakthrough that allows astronomers to study the fundamental components of stars and the possibility of forming planets earlier in the universe than thought.

The Pillars of Creation are located in the Eagle nebula, 6,500 light-years away. The pillar-like structures are filled with semitransparent gas and dust and are continuously evolving. They are the birthplace for young stars, some of which are seen ejecting material in wavy patterns. These stars are only a few hundred thousand years old and will continue forming for millions of years.

The theoretical physicist and mathematician Freeman Dyson (1923–2020) developed the concept for this sphere in 1960. He proposed searching the sky for anomalously large infrared signatures, possibly signaling their presence. Such a civilization would be so advanced that it could relocate its star with a stellar engine, an upscaled version of a Dyson sphere called a Shkadov thruster. This theoretical concept using gravity and the radiative force was introduced by Viorel Badescu and Richard Cathcart in 2000. Type III civilizations could theoretically achieve all of the above, but at a galactic scale, including harnessing the power of black holes, and having the ability to counteract cosmic expansion to maintain clusters of galaxies together for energy and power. Any Type II and

Illustration of the Dyson sphere concept surrounding a star and made up of independently orbiting panels.

Type III civilizations would collect so much information that they would have to broadcast almost permanently, thereby increasing the odds that Type I civilizations could intercept their emissions.

Although this may sound like science fiction, physicists, cosmologists, and astronomers conceived many of these ideas by projecting today's knowledge into the future. If anything, we are still a very young space-faring civilization with much to learn about how our universe works. As we keep making progress, these concepts could be just the tip of a cosmic iceberg of a reality far more wild than fiction. As an example, we might want to consider how the quantum revolution is changing our perspective of the universe we live in, from the subatomic level to the cosmological, demonstrating how we are all connected to everything every day.

Like the Drake equation, the Kardashev scale had many adaptations and extensions over the years. They include the addition of a Type 0 civilization, which extracts energy, information, and raw materials from primary organic-based sources and reaches advanced computing and orbital spaceflight. Humanity in its current stage is a Type 0 civilization and, consistent with its definition, vulnerable to extinction through natural disasters, exhaustion of its resources, and societal collapse. At the other end of this expanded scale, Types IV and V would be capable of harnessing the energy of the entire universe or even controlling collections of multiverses, and such a level of advancement would make them simply indistinguishable from the cosmic background.

In other variations on Kardashev scales, such as the classification proposed by the aerospace engineer Robert Zubrin, a Type I civilization has mastered all of its home planet's resources. Type II masters its solar system; Type III has access to the full potential of its galaxy. Unlike the Kardashev scale, this classification does not focus on energy usage. Instead, it emphasizes how widespread a civilization is in space. Each level is broken down into mature and immature types defined by their available technologies. Carl Sagan also proposed an alternative scale based on "Information Mastery." He assigned the letter A to represent 106 unique bits of information, each subsequent letter expressing an order of magnitude increase over the previous. A letter Z civilization would have 1,031 bits.

But, of all the scales, John Barrow's is probably the one that departs from all others. Instead of searching outward in space, he looked in the opposite direction. John Barrow (1952–2020), a cosmologist, theoretical physicist, and mathematician, proposed that advanced civilizations could manipulate their environment at increasingly small scales. In this fascinating concept, a Type I-minus civilization is capable of manipulating objects over the scales of themselves. Type II-minus can control genes. Type III-minus controls molecules. Type IV-minus controls atoms. Type V-minus controls protons. Type VI-minus controls elementary particles like quarks, and Type Omega-minus can manipulate the basic structure of space and time.

Now imagine just for a moment that some of these thought experiments reflect reality, that these civilizations exist out there in the depth of space, if only just a few, and you will never be able to watch the night sky in the same way again. The universe could be teeming with federations of galaxies, time travelers, stellar harvesters, cosmic dwellers, curious beings like us, born from the same universe, related to us by stardust, but separated by light-years. So then, in such a cosmic ocean of promises, how can we connect these pale blue dots together?

THE SETI PROJECTS

Making contact and detecting extraterrestrial technosignatures have been the objectives of the SETI search for over six decades now, since

Frank Drake, Giuseppe Cocconi, and Philip Morrison converged on the idea that scanning deep space to listen for nonrandom patterns of radio waves could provide a way to search for signs of alien civilizations. SETI is a scientific and technological endeavor conducted through targeted searches of nearby sun-like stars and systematic surveys scanning the sky in all directions. It accomplishes this across several domains, including radio astronomy and optical SETI, signal processing, citizen science, and active SETI, also called METI (Messaging Extraterrestrial Intelligence).

Historically, the first modern SETI project was Ozma in 1960, which Frank Drake carried out with the Green Bank Telescope in West Virginia. Subsequent projects included the High Resolution Microwave Survey (HRMS) selected by NASA, which would have scanned 10 million frequencies using radio telescopes, if the project had not been abruptly canceled following an effort led by Senator Richard Bryan to terminate NASA's funding for the SETI program.

A House-Senate conference committee subsequently approved the Senate plan, stopping the HRMS before it could start. Barney Oliver, a fervent supporter of SETI and head of research and development at the Hewlett-Packard Company, stepped up and was instrumental in organizing a fundraising campaign that gathered enough support to cover the cost of operations through June 1995. The now privately funded program run by the SETI Institute was aptly named Project Phoenix, and Jill Tarter, the project scientist for HRMS, became its director. From February 1995 to March 2004, Phoenix surveyed nearly eight hundred sun-like stars, one by one, within 240 light-years of Earth. Phoenix used the largest radio telescopes in the world at the time, including the Parkes Observatory in New South Wales, Australia; the National Radio Astronomy Observatory (NRAO) Green Bank Telescope in West Virginia; Arecibo in Puerto Rico; Woodbury in Georgia; and Jodrell Bank Observatory near Manchester, England.

Between 2005 and 2007, the SETI Institute built its own radio telescope facility, the Allen Telescope Array at the Hat Creek Radio Observatory (ATA/HCRO). The ATA is the first radio telescope array designed from the ground up to be used for SETI. It is located 480 kilometers north of Mountain View, where the SETI Institute has its headquarters in the San Francisco Bay Area. The array currently has 42 antennas built out of 350. It is used both for SETI searches and cutting-edge radio astronomy research, such as the study of fast radio bursts (FRBs). FRBs are enigmatic transient radio pulses caused by high-energy astrophysical processes of unknown origin(s). Each pulse lasts between one and a few milliseconds. A recent study suggests that magnetars, a type of neutron star with a very intense magnetic field, could be the source for at least some of them. At the ATA, research and development efforts include designing, testing, and implementing new

The ATA's radio telescope array at the Hat Creek Observatory is the first array designed exclusively with SETI research in mind, which, prior to its construction, depended on the availability of other antennas used for more conventional astronomical research. The antennas are located in the Cascade Range in Northern California, 480 kilometers from San Francisco Bay.

signal processing algorithms and new search technology and strategies incorporated into focused projects.

Currently, the SETI Institute is involved in several cutting-edge programs: One is the ATA/HCRO upgrade, which aims at upgrading the forty-two-antenna array with new technology to shift focus from tech build-out to star system observation within ten parsecs of Earth. Another is COSMIC SETI, a collaboration with NRAO to bring a state-of-the-art SETI search to the Very Large Array (VLA) in New Mexico, enabling 24/7 technosignature searches. The SETI Institute

Laser SETI cameras in Haleakala, Hawaii.

also collaborates with GNU Radio, a group of open-source software developers, to potentially revolutionize SETI receiving equipment and data analysis. CHIME is constructing a new radio telescope outrigger at the ATA/HCRO to accurately identify FRBs, promising scientific breakthroughs. Laser SETI capabilities are being developed, providing an optical tool to search for extraterrestrial biosignatures by monitoring the sky for laser flashes, covering a vast portion of the sky. The institute has collaborative projects on international radio telescopes like LOFAR in Europe, MWA in Australia, the Lovell Telescope in the UK, and FAST in China for radio and optical SETI searches. Shelley Wright, at UC San Diego, is leading NIROSETI, another recent SETI project, this time in the near-infrared that searches for optical signals in collaboration with several universities.

Among the most recent research programs is Breakthrough Listen, the most powerful, comprehensive, and intensive scientific search for technosignatures ever taken. Headquartered at UC Berkeley, in 2016 it initiated a ten-year survey of the 1 million closest stars, the nearest one hundred galaxies, the plane of the Milky Way galaxy, and the galactic center. The project is sponsored by the billionaire Yuri Milner's Breakthrough Prize Foundation. Breakthrough Listen also started a collaboration with scientists from the TESS team to search for signs of advanced extraterrestrial life. Thousands of new planets discovered by the TESS mission will be scanned by Breakthrough Listen partners worldwide in the coming years.

Also, since 2016, UCLA undergraduate and graduate students have used data from the Kepler mission to search for technosignatures with the Green Bank Telescope by targeting the Kepler field, sun-like stars, and the TRAPPIST-1 system. The same year, the FAST telescope began operations in China's Guizhou province. FAST is roughly twice as sensitive as the world's next-biggest single-dish radio telescope, and China is now implementing its Commensal Radio Astronomy FAST Survey (CRAFTS). As of 2021, FAST had already discovered 500 new pulsars and over 1,650 FRBs.

Signal and data processing are areas of SETI research and development that have experienced substantial progress over the years. SERENDIP (the Search for Extraterrestrial Radio Emissions from Nearby Developed Intelligent Populations) was one of the early projects headed by UC Berkeley's SETI Research Center. It is the brainchild of Dan Werthimer and evolved in time through several generations of systems. Previous radio SETI projects used dedicated supercomputers located at the telescope to complete the bulk data analysis. Rather than having its own observation program, SERENDIP analyzes deep-space radio telescope data while astronomers observe, basically eavesdropping during observations. The main focus of the project is signal processing. SERENDIP makes a broad search on many channels and a wide range of frequencies enabled by technological advancements, allowing a much bigger and more powerful search compared to the past. SERENDIP has been installed on antennas worldwide, beginning in 1998 with the telescope near Parkes in New South Wales. There it performed large scans of the sky without specific targeting. The current version of SERENDIP has been installed on the Chinese FAST telescope.

At Harvard University, the physicist Paul Horowitz has led microwave and optical SETI research for decades, including an all-sky optical SETI survey in collaboration with Berkeley and UC San Diego. Horowitz developed Suitcase SETI, a portable spectrum analyzer to search for SETI transmissions with a capacity of 131,000 narrowband channels. The system was installed in 1985 on the eighty-five-foot Harvard/Smithsonian telescope at the Oak Ridge Observatory in Harvard, Massachusetts. The program evolved over the decades, including the addition of a version deployed in Argentina to search the southern sky, and the project eventually innovated to receive 250 million simultaneous channels. Unfortunately, the twenty-six-meter radio telescope that this endeavor relied on was damaged by a storm, and operations stopped.

The public's fascination for the SETI search has given some of its programs a new dimension. In 1995, the computer scientist David Gedye and collaborator David Anderson proposed performing radio SETI using a virtual supercomputer composed of many distributed computers connected to the internet. That's how SETI@home was born. Launched in 1999, it ended in 2020 after becoming one of the most remarkable public participation projects ever undertaken, and a citizen project before they officially existed. Within months of its launch, more than 2.6 million people in 226 countries volunteered their spare processing power to parse the enormous volume of data generated by radio telescopes doing SETI and ran about 25 trillion calculations per second!

PROBING THE COSMIC OCEAN FOR SIGNALS

CASPER, the Center for Astronomy Signal Processing and Electronics Research, and the Berkeley Open Infrastructure for Network Computing (BOINC) are SETI@home spin-offs and SETI community projects. They offer a variety of new research disciplines in countries around the world. Since we do not know what extraterrestrial communication could sound like, the project combs the sky for several types of signals, searching for microsecond transient radio pulses.

Despite their differences, all these SETI projects have one common goal: intercepting a signal from an extraterrestrial civilization. SETI has traditionally relied on passive listening and watching through radio and optical methods, but these approaches are not the only ones. METI, which stands for Messaging Extraterrestrial Intelligence, is a nonprofit research organization that creates and transmits interstellar messages to communicate with extraterrestrial civilizations. METI's objectives are to target nearby stars, rethink the nature of interstellar messages to be sent into space by building an interdisciplinary community working on their design, and examine people's views of transmitting them. On October 16, 17, and 18, 2017, the organization sent a message to the Luyten's star, a red dwarf located twelve light-years from Earth, from a radio transmitter at the EISCAT research facility in Tromsø, Norway. Concurrently, METI performs more traditional SETI experiments with its optical observatory located in Panama, where it searches for laser pulses from advanced civilizations. Also known as active SETI, the idea of transmitting is a subject of controversy. The most significant criticism against METI relates to the Dark Forest theory. The underlying concept is that contact of any type is risky, and a species should avoid giving away its location and maintain radio silence. Stealth alien

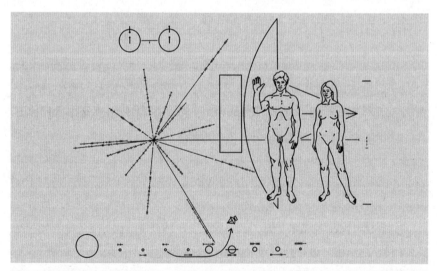

Before messages were sent by METI, humanity had transmitted its position in space through plaques designed by Carl Sagan and Frank Drake affixed to the Pioneer and Voyager probes, plaques that offer both humanity's greeting to other possible civilizations and the location of our planet from which these probes were launched.

civilizations could be lurking in the dark, according to the theory, and so we must tread with caution. This argument has often been used against METI. Some firmly oppose message transmission, including the late Stephen Hawking, fearing that this would reveal our position in space to malevolent alien civilizations.

It could be argued that we have broadcast radio signals unintentionally with our technology for over a century now. A few decades into the search, whether with SETI and more recently with METI, we still face the Great Silence. This could change one day, though, and the public often wonders what would happen if a message was intercepted. Some like to claim that this has already happened, but it is being kept a secret and, clearly here, the notion of secrecy is simply nonsense. The scientific method requires verifying data, implying that the discovery site would have to request confirmation from other observatories and astronomers in different parts of the world. In this case, secrecy is neither a possibility, as too many people would be involved very early, nor a policy because it would go against what science stands for.

Critically, what would be the reason to keep the most remarkable news in the history of humanity a secret, news that would be the answer to our ancestral question, Are we alone? Thinking that the public could not handle it is also one of the favorite themes of conspiracy theorists and for a good reason: it is one way to keep the conspiracy alive. It suggests that they know something that the public does not, which keeps the public interested and gives them the attention they crave. But what is most revealing is to see how those claiming that extraterrestrials with dreadful intentions are already here continue to post daily about it instead of running for cover. We often learn more by observing what does not happen than by listening to the background noise.

If a message is coming from many light-years away and extraterrestrials kindly announce themselves, they are probably not ill-intentioned and may just want to chat! If an evil species decided to conquer our planet, we would have to wonder why it had waited for so long to do so. Humans consider themselves the pinnacle of evolution on Earth. Yet, our species has lived in trees, caves, and huts and hunted with spears, bows, and arrows for hundreds of thousands of years. Extraterrestrials could have landed and taken over the planet then without even putting up a fight. Even if we assume that such species had become aware of us only recently and wanted to appropriate our planetary real estate, we would think that having the technology to traverse light-years in space would also mean they can land unceremoniously wherever they please. By then, it would be a little late for secrecy. The fact of the matter is that, for better or for worse, we attach human behaviors to extraterrestrials without any clear notion of who or what they are. Maybe they just don't care about us. Or, perhaps, we lack the clues of what type of life could be out there and how extraterrestrials might interact with us when that time comes. And maybe they already have and we did not recognize it.

Thinking about the potential social consequences of a SETI discovery was an early thought in the mind of John Billingham, a British physician, later the director of the SETI Program Office and director of the Life Science Division at NASA Ames Research Center. He organized discussions about this subject, keeping in mind the mistrust surrounding the Cold War years and the potential for misinterpretation of a signal by superpowers. He felt then that there should be at least some protocol to be taken if a signal was detected and confirmed. The first document produced in 1989 received limited attention but became the unofficial protocol. It was refined in 2010 by a subcommittee of the International Academy of Astronautics titled "Declaration of Principles Concerning the Conduct of the Search for Extraterrestrial Intelligence." The main guidelines expressed in the document are that a compelling signal should be verified and its extraterrestrial origin confirmed to avoid false positives from terrestrial interferences. Furthermore, "it should be announced to the world," and no reply should be transmitted without seeking international consultation.[1]

UAPS

All that precedes shows what the SETI search is and what it does. What SETI is not, however, is an endeavor involved in the investigation of unidentified flying objects (UFOs), now rebranded as unidentified aerial phenomena (UAPs). That often confuses the public (as indicated by the volume of emails I receive on this topic to my work email), but despite the confusion, the distinction and approaches between the two are clearcut. While UAP investigations involve evaluating surprising observations, SETI "conducts deliberate experiments and applies a series of tests to any received signal to verify that it truly originates from deep space and is not misidentified terrestrial interference."[2] In other words, SETI searches for evidence of extraterrestrials in their planetary and galactic habitats light-years away from Earth, while UAP investigations try to figure out whether extraterrestrials have entered our airspace. There are other fundamental differences between the two approaches. They can be summarized as the difference between scientific research based on systematic experiments, and police-style investigations trying to perform forensics on random observations.

We are all aware of the statistics on UAPs: About 95 percent of sightings are cases of mistaken identities easily dismissed as misinterpreted natural phenomena. Nearly 2 to 3 percent are likely to be secret military projects, such as the SR-71 in the past, and more advanced airplane technologies and the occasional Chinese spy balloons in present days.

Another percent or more will be attributed in the future to natural phenomena for which we do not yet have an explanation. In that respect, our planet's changing environment may contribute to its fair share of novelties in the years to come and possibly bring spikes in the number of cases.

But there are also those incidents that do not seem to fit into any of these categories. As an astrobiologist and a citizen, I am interested in that fraction of cases because, through their study, we will gain new knowledge by understanding what they are regardless of their nature. I therefore followed with interest recent reports and task forces, such as the latest Unidentified Aerial Phenomena Task Force (UAPTF) report on UAPs by the U.S. military and intelligence community relating to cases that occurred between 2001 and 2004. The document focuses on 144 incidents witnessed firsthand by military aviators and collected from reliable systems. Despite the formalized process, one of the document's main conclusions is that the available reporting remains largely inconclusive, and limited data leave most UAPs unexplained. Some patterns seem to emerge regarding their shape, size, and propulsion, and, of the 144 incidents, eighteen appear to demonstrate advanced technology.[3] The document states that evidence recorded across multiple sensors increases the chances that they are actual physical objects, some of them exhibiting unusual flight characteristics, although possible errors from the sensors are not entirely dismissed. The report concludes on multiple types of UAPs, listing five possible explanations: airborne clutter, natural atmospheric phenomena, U.S. government or U.S. industry developmental programs, foreign adversary systems, and a category labeled "other" for anything that does not fit the previous descriptions. Finally, it raises the possibility of a flight safety issue and a challenge to U.S. national security as potential evidence of breakthrough or disruptive technology developed by adversaries.

Following the report, the U.S. Congress moved to establish a formal office to carry out a coordinated effort to collect and analyze data related to UAPs. "Keeping the strategic edge and the nation safe" led to the inclusion of Senator Kirsten Gillibrand's UAP amendment in the National Defense Authorization Act, signed by President Joe Biden on December 27, 2021. As a result, a new office within the Pentagon called the Airborne Object Identification and Management Synchronization Group will now be tasked to implement a plan to test scientific theories related to UAP characteristics and performances.

Adding to government-led efforts, an initiative called the Galileo Project, spearheaded by Avi Loeb—the Harvard astronomer who proposed that the interstellar object 'Oumuamua is an alien light sail—will search for extraterrestrial technology near Earth. One of its objectives is to identify the nature of interstellar objects that do not resemble comets or asteroids; the other is to target UAPs. The Galileo Project includes over one hundred researchers, a telescope system at the Harvard College Observatory, and the use of AI to differentiate UAPs from natural phenomena. This project is just one of many, bringing hope that coordinated efforts could produce big dividends in the form of vast amounts of reliable physical data that will be scientifically analyzed.

This new approach may draw in more scientists who were reluctant to join an endeavor often officially dismissed in the past, including by the military. Past efforts were ridiculed mainly because of the unfortunate confusion between the academic research for extraterrestrial technosignatures in the universe, the study of accounts undertaken by serious UAP/UFO groups, and the folklore surrounding UFO fanatics. Paraphrasing another famous quote from Sagan, as a scientist, "I do not want to believe. I want to know." It is precisely what separates modern UFO gurus and their corteges of worshippers from scientific investigators. Over time, the influence of these gurus has undermined and unnecessarily stigmatized the effort of scientists who seek to bring attention to a research domain that is at the core of humanity's most profound question. Although this might seem like a harmless distraction, consequences are real in the collective psyche of a society in which science literacy is low and where scientific evidence is often denied and ridiculed at the highest level. All of this only encourages more conspiracy theories and reinforces their influence over the public, altering people's ability to think clearly, and ultimately distracting from the search for factual evidence.

This psychological manipulation develops into narratives where facts and fiction melt into each other like ice in summer. Fortunately, a fairly simple smell test separates authentic research from baloney. Authentic research draws its conclusions from the methodic analysis and verification of data and testable hypotheses; baloney bends backward to make data fit a predetermined conclusion. Alternatively, bad faith arguments will start with a "what if" proposition, which is then used as sole evidence for an inevitable conclusion.

Regardless of the origin of the sightings, the report's release and the new initiatives bring a chance to collect fundamental knowledge about UAPs. As the scientific method is increasingly considered in this new environment, there might be more room for partnerships between government agencies and private research institutions. For example, on the day the UAP report was released, NASA posted a FAQ titled "Unidentified Aerial Phenomena (UAPs/UFOs)"[4] that clarifies NASA's focus, and points out that it shares its data with the public. "[NASA] does not actively search for UAPs. However, through Earth-observing satellites, NASA collects extensive data about Earth's atmosphere, often in collaboration with the world's other space agencies. While these data are not specifically collected to identify UAPs or alien technosignatures, they are publicly available and anyone may use them to search the atmosphere."

NASA's press release concluded: "If we learn from UAPs, it would open the door to new science questions to explore. Atmospheric scientists, aerospace experts, and other scientists could all contribute to understanding the nature of the phenomenon. Exploring the unknown in space is at the heart of who we are."

On the same day, the SETI Institute posted a press release[5] that explained "if the pending government report indicates that there is a possibility that at least some UAPs might, indeed, be of extraterrestrial origin, then perhaps there will be an effort to pursue their study using the precepts of well-designed science experiments."

Moreover, and contrary to what has been alleged over the years by conspiracy theorists, SETI researchers do not conceal information about UAPs out of some fear that it would make their research obsolete—or for any other reason. They simply do not have any information on a domain that is not the focus of their research. We would actually welcome the confirmation, if it ever arrives, that some of these unidentified objects are of extraterrestrial origin. SETI is a scientific endeavor, not an industry tasked to deliver a finite product at a given deadline. And, far from ending the journey for scientists, the discovery of an alien presence in our planet's airspace would allow SETI to open a limitless horizon of new science questions to explore. And finally, simple common sense should tell us that if aliens visited Earth in the past or are still visiting, they must master extraordinary technologies. They are unlikely to have made the trip just to stack stones in the shape of pyramids to leave us scratching our heads over cryptic messages.

On the other hand, why are there pyramids of comparable proportions all over our planet? Why are patterns and geometries repeated across the Earth, why are there common narratives shared by cultures that did not interact? These are only a few of many truly fascinating questions that, as a scientist, I am curious about and willing to explore.

The conclusions might or might not necessarily lead to extraterrestrials, but alternate hypotheses could be equally mind-shattering if verified. It may be that through time and art, in our daily lives and heritage, all humans and all life on Earth express what life is in a language that transcends cultures and distances, a blueprint deeply rooted in all of us, individually and collectively. Whether this inner language exists, is unique to Earth, or is a universal cipher, it may end up being an essential key to decoding alien intelligence one day.

12

PARADOXES, PARADIGMS, AND THE GRAMMAR OF LIFE

Our generation is living through a truly unique moment in time when humanity finally has just enough knowledge and technology to begin searching for life beyond Earth. The past decades saw a relentless succession of extraordinary discoveries, a time when our eyes opened to the vastness of our universal citizenship. And, just like children on their first adventure, we marvel with every step we take. If we are not so arrogant as to annihilate ourselves, we might finally make it through the narrow gates of teenage civilization to enter adulthood. Then who knows what cosmic horizons will be out there for us to explore? Who can tell how many others undertook the same journey light-years away from us, and how much knowledge we could share on the day we finally meet? Wherever this cosmic voyage takes us, there will be no other times like today, when everything was new and untouched, and when we laid eyes on landscapes in our planetary neighborhood for the very first time. Our generation pioneers the opening chapter of this nascent spacefaring exploration. We can only hope that many more will follow, and countless books of epics will one day be placed under our name in the collective cosmic library of life.

SANDWICH IN HAND

But for now, this journey is still full of paradoxes. I often compare it to my all-time favorite comic strip, *Calvin and Hobbes,* by the American cartoonist Bill Watterson. Calvin is a mischievous little boy who always shares wisdom for the ages, like this one: "Sometimes I think the surest sign that intelligent life exists elsewhere in the universe is that none of it has tried to contact us." Calvin lives in a world where everything is new and sounds and looks magical, and where everything can happen. I have a particular fondness for those strips where he leaves home with his stuffed tiger, Hobbes, to conquer the universe before nightfall and packs a sandwich for the journey. I think this is the perfect metaphor for where we are right now in our search for life in the universe. We are bursting with enthusiasm and expectations for what's out there to

The International Space Station (ISS) plays a crucial role in maintaining humanity's continuous presence in Earth's orbit. It is a unique laboratory that offers invaluable insights into life in space, and
serves as a stepping stone toward a future where mankind ventures beyond Earth. The ISS was recently joined by the Chinese station Tiangong (Heavenly Palace).

discover, but we have no real idea of the scope of the endeavor, what is needed for it, or where it will lead us. We simply know that we have to set the journey in motion. That's our first step, and just like Calvin, we go as far as our sandwich can take us (in our case, knowledge).

As we set out to discover life beyond our planet, one of the greatest paradoxes of astrobiology is the lack of a consensus definition for what life is. And this is just one of many paradoxes. We are also still pondering over life's emergence, wondering if it is a single event or a gradual transition from the non-living to the living, or whether what we can understand today is only the visible manifestation of a more profound universal process. And since these questions would not be complex enough on their own, artificial life has recently come to complicate the matter further. Our journey of exploration is thus all at once the means, the goal, and the learning tool, and this has implications for our approach.

We use our planet as the standard and we understand reasonably well where prebiotic chemistry and biochemistry can take place, but our conception of life's emergence is essentially based on our comprehension of its building blocks and environments possibly favorable to its inception. This approach has guided us in the past few decades as we characterized habitability in the solar system, and this rationale is still present in the current missions. We "follow the water" on Mars in ancient lakes or search for minerals that testify to an aqueous past like in Meridiani Planum. There, we also seek the presence of ancient sources of energy and carbon, physical shelters, and the elemental bricks of life. Ocean worlds excite us because their environments resemble those we believe led to the origin of life on our planet. In terrestrial analogs to these planetary environments, we set out to explore where life could have emerged on these worlds and how its signatures may have been preserved.

We do not search for extraterrestrial intelligence. Just as we do not know what life is, we do not really know what intelligence is either.

Instead, we try to eavesdrop for evidence of its technology in coded radio and light signals traveling through interstellar space or in the signatures of alien megastructures. It is a step that actually makes a lot of sense at this point in time. Considering the light-years separating us from our closest potential neighbors, and how complex and hazardous space travel may be for fragile biological life-forms, many think that we are unlikely to meet E.T. in person anytime soon. Chances are that the first contact will be through pioneer extraterrestrial robots, possibly resembling the self-replicating von Neumann probes.

Whether through the notion of habitable environments or extraterrestrial technosignatures, we do not search for life per se, but for indirect evidence of its presence. It is an exploration by approximation, one that may accumulate converging evidence of life's presence, but may leave us short of an unambiguous response for years to come, as demonstrated a few times already with the Viking experiment data, the alleged fossils in the Martian meteorite Allan Hills 84001, and most recently, phosphine in the atmosphere of Venus. But now that planetary exploration is transitioning from the study of habitability to the search for life, the time has come to develop a strategy that could give us the necessary physical, chemical, and biological lines of evidence to help us conclude with some certainty that life was, indeed, discovered.

THE PLUMBING OF LIFE

With that in mind, and to prevent the proliferation of false positives and false negatives, astrobiologists recently developed a "Ladder of Life Detection."[1] This ladder is an eight-step method that provides a set of converging evidence augmenting confidence that life was discovered. On it, life detection is defined as "searching for past or present biosignatures in the form of complex organic matter that cannot be formed only abiotically, or concentrations of elements required for life as we know it and establishing the context they are found in."

By definition, biosignatures are any measurable phenomena indicating the presence of life, and they are divided into various categories. They can be objects, substances, or patterns whose origin specifically requires a biological agent to form. Because of the paradox we face with the lack of a consensus definition for life, the "primary value of a biosignature is not the probability that it was made by life, but rather the improbability that it was made by non-biological processes" (the astrobiologist David Des Marais). Environmental context is thus critical to characterize first the non-living chemistry of the worlds we explore and the setting from which the samples originate. For example, contextual information can be provided by rocks, specific mineralogy, and environmental and climate history.

The Ladder of Life Detection is time-stamped to our current technology and instrument sensitivity, and to our understanding of physicochemical and biochemical processes. In other words, it is a tool in progress that will improve as knowledge advances. For now, its eight rungs include instrument sensitivity, the lack of contamination, and repeatability of measurements. In addition, one or more features must be sufficiently detectable, preserved, and reliable. In this context, they can be differentiated from a non-living signal, compatible with life as we know it, and biological interpretation must be the last-resort hypothesis. The various rungs of the ladder lead to increasingly compelling indications of biology. Evidence includes biofabrics (rock and soil textures created by the presence of life), metabolism and metabolic by-products, biomolecule components, functional molecules and structures, growth and reproduction, and signs of adaptation and selective pressure (Darwinian evolution).

This is astrobiology's current intellectual framework, and this approach embeds a set of fundamental assumptions, which are the foundation of our current strategies to search for life in the solar system and beyond. In it, living and non-living things are clearly separated. Biosignatures are features only created by life. Most of everything else is characterized as nonbiological processes. Sometimes, we stumble into ambiguous features that current knowledge and technology cannot assign to these categories. Viruses are an example of such ambiguity.

Alexander Oparin himself did not see a fundamental difference between a living organism and lifeless matter. For him, life's characteristics arose as part of the evolution of matter, and prebiotic chemistry transitioned into biology. But even this apparently simple statement is far from being as straightforward as it seems. It is still at the heart of debates and unresolved questions surrounding the origin of life, and chiefly, what could be the nature of this transition. For the physicist Eric Smith and the biophysicist Harold Morowitz, it is "a necessary cascade of non-equilibrium phase transitions that opened new channels for chemical energy flow on Earth."[2] Others suggest that it really depends on our perspective. For instance, if we reduce life to a single living compound, then it can be well-defined by the appearance of the first replicating molecule. But if we think of life as a stepwise evolutionary transition, then it might be meaningless to draw a strict line between living and non-living things.[3]

Regardless of the perspective we choose, we must explain the apparent selectivity of biochemistry and prebiotic chemistry. For instance, why did life use specific subsets of molecules rather than others? In other words, how did prebiotic organics narrow down to today's chemistry of life, and why did some molecules lead to life, whereas others did not? An answer to the first part of this question seems to be coming into greater focus today. Life may have used specific raw material simply because they were the most abundant and easy to form in the early Earth environment, as suggested by the formation of nucleic acid precursors starting with just hydrogen cyanide, hydrogen sulfide, and ultraviolet light, and the conditions that produced these precursors were also right to create the raw material to make amino acids and lipids.

If this scenario is correct,[4] then a single set of reactions could have produced most building blocks of life simultaneously. Further, we neither know nor understand the mechanisms that led to the production of prebiotic molecules in various environments or how critical these environments were for the stability and accumulation of organic molecules. Decades of research and exploration have clearly demonstrated that organic molecules are produced in abundance in the interstellar medium and planetary worlds of all sizes, but how much was delivered to the Earth during accretion? How much was manufactured on our planet, and which proportions of those led to life? This is unresolved. Other outstanding questions relate to the role of various energy sources in producing different organics. These questions represent only a handful of many fundamental interrogations that remain to be addressed before a clear picture can emerge.[5]

While the bricks of life may have been abundant and easy to assemble, the significance of chance, of chaos, still cannot be entirely discounted. How much, and when did it come into play in that process? Random events of various origins could have acted individually or in concert to set prebiotic chemistry on its path to biology, but research shows that it is not a necessary condition for an emergence of life. These random events definitely altered life's evolution repeatedly after it had taken hold, including through asteroid and comet collisions, climate events that drastically altered environmental conditions, or geological or environmental catastrophes, to name a few. This notion that random events shaped the history of life on Earth, from emergence to its evolution, was already proposed by the biochemist Jacques Monod[6] in the

1970s and is a central argument of the Rare Earth hypothesis, driving the conclusion that simple life must be abundant in the universe, but not complex intelligent life. Since then, the staggering number of exoplanets discovered by Kepler and TESS has considerably weakened this argument. In Kepler's universe, randomness is likely to determine what type of complex life will emerge, but unlikely to limit its abundance.

Solving these questions is an essential part of what astrobiology seeks to achieve. Otherwise, if we cannot draw the boundary between living and non-living on our own planet, we will continue to struggle to recognize evidence of life on other worlds. And that's only for life as we know it. We must also tackle "nonstandard" life questions, including alternative biochemistries or different habitability constraints that may have emerged on other worlds. Addressing the search and detection of nonstandard life requires identifying and characterizing what attributes are unique to life on our planet and which ones could be universal. Meanwhile, this distinction may be difficult to make until we find another type of life beyond Earth. Yet, some attributes seem reasonable candidates to enable essential universal functions. For instance, life harvests energy and converts it to operate. It has physiological functions and sustains metabolism to maintain steady internal physical and chemical conditions, growth, and self-replication. Life apparently needs sets of specific environments and elemental components to achieve assembly, to function, and to self-perpetuate.[7] Extending these questions to complex and technologically advanced life, we must learn how to infer the possible evidence broadcast in space by a distant, technologically advanced civilization from the current influence of human activity on its own biosphere.[8]

All these questions relate to life's origin and its interaction with the environment at various stages of its evolution. It is about how life assembles, what life does, and the evidence it leaves behind. This intellectual framework is what I personally call "the plumbing of life." In it, *what life does defines what life is.* The working definition for life for NASA's planetary and space exploration is "a self-sustaining chemical system capable of Darwinian evolution." But this is only one of the many definitions from philosophy, religion, and science, and more are being proposed today. The difficulty in reaching a consensus is due in great part to the fact that life is a process, not a substance. Likewise, new scientific paradigms have been proposed in the past two decades, which do not consider the "plumbing of life" as what defines it. Instead, they suggest that it is only a manifestation of what life is. In other words, *explaining the origin of life might not be enough to define it.*[9]

The working definition used for planetary and space exploration implies physiological functions such as homeostasis, a relative state of equilibrium between interdependent elements, organization, metabolism, growth, adaptation, response to stimuli, and reproduction. It also includes a cell's ability to receive, process, and transmit signals within itself and the outside environment. Other modern scientific definitions of life can be arranged in several categories: biology, biophysics, and theories of life as simple and complex living systems. Recent developments in AI and machine learning have led to considering artificial life as a possible new form of life, although this is disputed. Meanwhile, science and biological engineering have opened up the universe of synthetic biology.

LOOKING FOR DEFINITIONS

In trying to define life, the work performed by the molecular biophysicist Edward Trifonov in 2011[10] is probably one of the most original approaches. Using 123 definitions of life from science, history, and philosophy, he completed a statistical analysis of word occurrences and reduced them to a "minimalistic definition." Intermediary steps removed connective words and synonyms. Next, the most commonly occurring words, such as *life, system, matter, chemical, complexity, reproduction, evolution, environment, energy*, and *ability*, were listed. Trifonov then created a more concise definition by removing terms implicitly involving each other. For instance, *metabolism* implies both energy and material supply, which also represents environment. The final reduction in terms led to the two independent notions of *self-reproduction* and *changes*. Interestingly, these are two contradictory terms: self-reproduction implies an exact copy, which somewhat excludes Darwinian evolution (changes). To resolve this paradox and logically combine these terms, Trifonov added a third notion. Ultimately, his minimalistic definition of life became: "An almost exact self-reproduction with variations," a definition that approximates Oparin's statement that "any system capable of replication and mutation is alive."

In that perspective, the 2022 experiment performed by the University of Tokyo researchers of self-replicating DNA that diversifies and develops complexity demonstrated an example of transition from simple chemical systems into complexity following Darwinian evolution.[11] It seems, then, that the empirical approach would support the idea that *life is what life does*, and there might not be much more to it (so to speak). At least, there is enough to go by with this definition and use it as a working hypothesis in the search for life beyond Earth. But not everybody agrees.

While he acknowledged that Trifonov could be close to a solution, the molecular biologist Ernesto Di Mauro rejects the idea that the minimalistic definition—or any definition proposed thus far—actually defines life. Instead, he proposes that all we have now are *descriptions* of life. We could illustrate this point by saying that defining life by describing it is the same as saying that we can define the atmosphere by describing a bird flying in the sky. If this is true, then we can certainly expect our description of life to improve with scientific and technological advancements, but it might not get us any closer to a definition. Hence, the following question: Does understanding what life is and defining its nature require a change in perspective?

In 1944, the physicist Erwin Schrödinger had proposed to look at life from the perspective of physics instead, which led to the development of several biophysical definitions. For Schrödinger, living matter is what "avoids the decay into equilibrium"—or at least delays it by fighting entropy.[12] A more recent version of a biophysical theory of life was proposed by the physicist Jeremy England in 2013, who used statistical physics to explain the spontaneous emergence of life as an inevitable outcome of thermodynamics.[13] Aspects of his theory have been successfully tested in computer simulations since. They show that groups of random molecules can self-organize to absorb and dissipate heat from the environment, leading to the rise of entropy (molecular disorder) in the universe. England calls this restructuring effect the "dissipation-driven adaptation," an effect that fosters the growth of complex structures, including living things. The living systems theories approach the same aspects but, as the name suggests, from a systemic perspective, and consider the "reciprocal influences of environmental and biological fluxes"[14] (coevolution) at the level of ecosystems rather than the level of individual organisms. Several variants of the living systems theories exist, but the best known is probably Gaia.

Gaia proposes a symbiotic relationship between life and its environment, in which they evolve together as a single, self-regulating system maintaining the conditions for life on Earth.[15] The scientific foundation of Gaia lies in the observation of how the biosphere and life's evolution contribute to maintaining environmental factors of habitability in a preferred and relatively stable equilibrium. The notion of Gaia was first articulated in 1785 by the Scottish naturalist, also known as the founder of modern geology, James Hutton. The chemist James Lovelock and the microbiologist Lynn Margulis codeveloped its modern version in 1974. While the notion of coevolution of life and its environment was already known, the uniqueness of Gaia was to show that equilibrium is actively pursued to maintain conditions optimal for life. The idea was not widely accepted when first proposed, but some predictions were verified, and experiments have supported some of its aspects since. It is now recognized as a theory, and it is studied in earth sciences, biogeochemistry, and systems ecology, where it provides a systemic view on the consequences of climate change.

Taking a different approach, Chris Kempes and David Krakauer's "Multiple Paths to Multiple Life"[16] makes the radical claim that a new theoretical framework should reveal many origins and many types of life. Here, the various origins of life do not relate to the plausible successions of emergences and extinctions suspected to have happened in the hostile environment of early Earth. Instead, life would have evolved not just once but multiple times independently. They argue that a new theory is necessary to recognize life's full range of forms, and their three-part model considers "the full space of the material in which life is possible; the constraints that would limit the universe of possible life; and the optimization processes that drive adaptation." Some adaptations are perceived as new forms of life instead, and their definition of life is broadened to include "cultures that feed on the material of minds, [including] computation," and much more. It is an interesting concept that may offer a possible bridge between the living systems theories and those of England's and Gaia. It also provides space to consider artificial life as being real life. And there, not everybody necessarily agrees, either.

ARTIFICIAL *AND* ALIVE?

With the definition of natural life being so elusive, it is not unexpected to find artificial life at the heart of a controversy that focuses on the rapid development and expansion of synthetic biology, nanoscopic robots (nanobots), and artificial (intelligent) systems. Each passing day, advanced technologies are increasingly blurring the line separating living from non-living. With artificial intelligence, one can still argue that the substrate of intelligent systems is not natural and manufactured by biological entities (us). This being considered, although it is an engineered second-order emergence, could artificial life still be (or become) alive instead of just being the synthesis and simulation of biological systems? Here, opinions diverge.

Today, we are probably a few decades away from a human-level artificial intelligence. AI is definitely capable of completing tasks much faster than humans, such as analyzing billions of data points per second, for instance. But AI systems are dedicated to particular problems, and as long as humans train them, many are not willing to consider them as anything other than highly capable tools. But what will happen when AI is not subject-specific anymore and capable of acting in any domain? This is where most of the current controversy resides, in particular around the concept of an "artificial intelligence singularity," a point sometime in the future when intelligent machines will exceed the cognitive intelligence of humanity (in other words, become a lot smarter than us), and possibly develop traits of life, intelligence, and awareness.

This concept implies (technological) evolution, and artificial intelligence is progressing constantly. And with it, so are domains such as evolutionary robotics, an approach that uses Darwinian principles of natural selection to design robots; autonomy, which is the ability for an AI system to make decisions without human interference; self-organization, a key theory in developmental neuroscience adaptation used in artificial neural networks for pattern recognition and classification mechanisms; and the list goes on.

If these terms sound familiar, that's because they are for the most part analogous to language used in life sciences, neuroscience, and evolutionary biology. And the reason is not all that surprising. Humans are building AI systems and intelligent robots following their own familiar blueprint—and to some extent in the same fashion as we are searching for life as we know it in the universe. This is what we know and what works. But at this point in our technological development, despite the level of anthropomorphism we bring into it and how much it can accomplish, the vast majority of experts would agree that AI technology is not alive, but just an extension of human activity. On the flip side of the argument, it could also be said that AI does not necessarily have to duplicate humans to possibly become alive one day.

Not surprisingly, either, opinions diverge over what it would take for artificial life to be considered alive. Conditions could include its ability to take care of itself and its species, the aptitude to repair and upgrade itself and adapt to its environment, as well as an awareness of its own goals. The replacement of parts of itself—or even its whole—could be perceived as a substitution to reproduction in natural systems, and improvement as an analog to adaptation. AI could also generate new algorithms toward endless self-improvement as an analogy to Darwinian evolution. When it reaches that stage, AI will not be a mere extension of humans anymore. It will produce independent, super-intelligent, and more capable entities that may also diverge from a completely anthropomorphic blueprint and develop their own forms of life and intelligence. Yet, would that give them an identity or self-awareness, or a sense of purpose? For some, this is a bridge that they are not willing to cross just yet. Maybe AI will become alive and evolve beyond its original human blueprint.[17]

ARTIFICIAL INTELLIGENCE, TRANSHUMANISM, AND ETHICAL DILEMMAS

The evolution of AI toward increased complexity and capabilities raises profound ethical issues, in particular regarding the future of the relationships between humans and superintelligent machines. However, if we want to caricature the basic components of AI, it is pure technology: chips, transistors, connectors, wires, nuts, bolts, and a fresh coat of paint. The intelligence is all in the assembly. But this is somewhat different with synthetic life-forms now capable of self-replicating, such as xenobots. These nanoscopic robots (nanobots) are designed to support regenerative medicine or monitor the environment, among other things.

For now, they are collections of cells that can be programmed to corral other cells. Since they are made from genetically unmodified frog skin and heart muscle cells assembled into nanoscopic balls, they are living organisms. But they can also be programmed, which makes them robots. And apparently, they have a mind of their own. The team that built them discovered that they could swim, find single cells in their petri dish, and assemble hundreds of these cells to create "baby xenobots." They carry them in their "mouth" for a few days like tiny Pac-Men until they are fully grown, look like their "parents," and start assembling cells on their own.[18] In other words, these nanobots have begun replicating without anybody asking them to. They work together to create something far different from their initial frog's genome, and while they fulfill some of the criteria of living organisms and are partly biological, they are also part robots. The question then becomes: Are they living or non-living?

There, too, synthetic biology is filled with gray areas and ethical issues, mainly because of the creation of entities that might possibly fall in between the descriptions of living and non-living. In 2017, the physicist

Louis Del Monte predicted that nanobots would represent a second technological singularity. He defined this as the development of the first artificial life-forms with enough intelligence to carry out programmed functions and reproduce.[19] Because of the symbiotic relationship between nanotechnology and artificial intelligence, the advent of this second technological singularity was, according to Del Monte, dependent on the first one (AI exceeding the cognitive intelligence of humanity). He predicted the first singularity to occur in the 2040s, but apparently, the nanobots could not wait.

The merging between AI and natural life is real and happening at all scales and at various degrees of integration, from pure AI to synthetic biological systems, and from nanobots to the most advanced machines. The next step up in that chain of evolution is transhumanism, which involves the development of human-enhancement technologies designed to increase sensory reception and cognitive capacity, improve health, and extend life. In one of its futuristic visions, transhumanism could lead to radically enhanced humans akin to cyborgs, bypassing natural evolution to replace it with technology. In another, it brings the emergence of greater-than-human machine intelligence, allowing the merging of human and machine consciousness.

Leaving aside what cognitive consciousness could mean for robots and what we may think of a cyborg humanity, we can focus on the reasons that would push our species to consider altering its ancestral biological substrate for the first time by merging it with an unnatural and engineered one. This is a completely different evolutionary path that, if completed, could produce a new type of humanity altogether. It is a risky path as well, since the process involves the creation of technological (AI) entities that may, one day, become competitors on our own ancestral turf, while being a lot smarter and more capable than we are. The answer as to why we are taking such a unique evolutionary step has to be found in the risks-versus-benefits, which tells us something very profound about the nature of life, what life actually is.

Practically, there could be a true evolutionary edge in allowing the creation of hybrid and more robust versions of ourselves. Besides going beyond physical and intellectual limitations, it could allow our species to expand its reach in, and connection to, the physical universe. In particular, it could provide more protection and would facilitate our expansion in space and any other extreme environments that are not indigenous to humans. Ultimately, it could give access to new environments, which multiplies the odds of preserving at least some of our version of life. In a different way, and on a different substrate, the same idea of life's preservation is also present in current initiatives collecting human and animal DNA, which is preserved in secured places on Earth, and soon on the moon, "to safeguard life," or more specifically here, the genetic encoding information of Earth's life.

Artificial intelligence, virtual worlds, and digital transformation further enable the expansion of our sensory perceptions into realms located beyond our day-to-day physical reality. Something like the Neuralink implant may reach a point where it could connect the human brain with computer interfaces via artificial intelligence. While its first applications may be the medical support of paralyzed patients, it may also be used one day to enhance the human brain and lead to a symbiosis with AI. In a different, albeit related domain, AI-driven virtual worlds and avatars give us access to inner universes of limitless dimensions where we can immerse ourselves into simulated experiences. Although these experiences are virtual, the stimuli to our brains are real. Research shows that one of the effects of virtual reality is to boost brain activity that is critical to learning and memory and to enhance neural connections, allowing us to learn faster and to retain information more effectively.

IT IS NOT BY CHANCE

All these different tools, technologies, and transformations have something in common that reveals the nature of life. They show that what life does is *not* what life is. It is more of a means to an end, and the nature of life resides in what is ultimately achieved in the process. Life acts as if it is a process that fundamentally aims to improve and broaden our experience of being, of our awareness of being alive, by multiplying the sources of stimuli and information, the interactions and connections with the environment—whether this environment is physical, mental, or spiritual, real or virtual—and to seek an ever-increasing complexity to ultimately give more dimensions to our consciousness and to the universe. Our own path as a species and its current crossroad with AI suggests that the process of life does not care so much about what materials it uses to achieve a goal; it simply cares that the goal is achieved.

Whether the universe is fine-tuned to allow life and consciousness rather than coming about at random[20] has yet to be demonstrated, but there seems to be very little randomness in how the process of life manifests and organizes itself in what could be defined as the *grammar of life*. Its structure, patterns, and signatures are everywhere in and around us and deeply wired in all life on Earth. We saw that the emergence of life starts with the organization of individual atoms, then simple and complex inorganic molecules, complex organic molecules, replication, and increasing biological complexity. We find the same structure in our languages in striking parallels.

Think of letters as individual atoms that, brought together, create syllables and then meaningful words. By themselves, words are descriptive of what they are, but it takes verbs to assign actions or states. Syntax arranges words to make sense of the information conveyed in a sentence, in the same way DNA organizes our genetic information to make sense of it. Complements and conjunctions add depth, information, and complexity. Languages can be very different but, ultimately, they all reflect the same goal of forming concepts and thoughts, gaining and passing on information, connecting, and facilitating exchanges effectively with the outside world or within ourselves. In the biological realm, similar functions are achieved by exchanges within a cell and between a cell and its environment.

For humans in societies and civilizations, the spatiotemporal evolution toward greater diversity and complexity in the distribution of information (volume, storage, and transfer) has been realized so far through the invention of languages and alphabets, the printing press, the internet, and online management of information, and more recently, quantum computing, and the encoding of DNA with external, non-biological data. At grand scales, we find the same patterns of connectivity and increased complexity in the cities we build, which evolve in time from isolated outposts to megacities organized in increasingly effective trade and communication routes. At microscale, it is observed in the most desolate places on Earth with microbial oases in extreme environments. They grow in size and complexity following very specific patterns optimizing information and resource flow. And these patterns are observed everywhere in nature. Is this blueprint unique to life on Earth, and did it become nature's most successful and effective strategy over time through trial and error? Is this a universal blueprint? One thing is certain, though. The laws of physics and chemistry are universal and the building blocks of life on Earth are abundant and common, and although they might not be exactly the same elsewhere, the odds suggest that many more analog blueprints of the process of life could exist in the universe, in the same way synonyms provide different means to convey the same information in grammar. Now we just have to figure out ways to test this hypothesis and see how it may help us open new avenues to search for life beyond our planet.

This process of life that seems to emerge in our explorations reminds me of what Sagan said once while referring to the star stuff we are made of, and what he called the cosmos within: "We are a way for the universe to know itself." I would just add that, judging from what we have learned so far, it seems that we are also a way for the universe to become ever more alive, aware, and conscious.

EPILOGUE

TIME AS A COSMIC MIRROR

"The nature of life on Earth and the search for life elsewhere are two sides of the same question—the search for who we are."
—Carl Sagan

B ut who are we, Carl?

This question comes back relentless and more urgent in my mind each passing day, as we ignore the signs of our changing planet, the bloodred skies filled with the soot of burning forests, marine life suffocated by the onslaught of plastics pollution down to the bottom of the Mariana Trench, glaciers vanishing, polar caps melting, droughts driving us to choose between water and electricity, polluted waterways, climate-driven exodus and wars, and the rise of fascism's ugly face once again. The answer to who we are at this moment in time is as much about understanding the origin and nature of life in the universe as it is to look at ourselves in a mirror and decide whether we are destined to last as a civilization or to vanish before having a chance to express all the good there is in us, our love and compassion, our art and ingenuity, and our boundless curiosity. We have come to the crossroads of time and must decide about our future as a species and as a civilization.

PATHWAYS AND DESTINIES

On May 13, 1897, the first radio transmission left our planet. At the same moment, a cosmic hourglass started to count away the years our civilization would remain detectable, the (L) factor in the Drake equation, the last factor to the right. Then the question became whether humanity could be a long-lasting flame spreading across the galaxy, steadily climbing the rungs of the Kardashev scale, or a fleeting spark of brilliance soon forgotten, like so many doomed evolutionary pathways before us. (L) involves much more than just a duration. It is a time of awakening and awareness. It is a time when a civilization becomes capable of altering the natural course of its planet and changes its relationship with its environment. It is likely a cosmic initiation that all young technological civilizations face at some point in their evolution, a rite of passage when the true value of life at a planetary scale is being branded into their souls. It is a make-or-break moment and, unlike previous thresholds in evolution, it is self-made.

Our own passage into cosmic citizenship was accompanied by winds of change. In the 1950s, the human population exploded through exponential growth that has not abated since. Although our influence on the climate became obvious in the industrial era,[1] the first signs of global warming were detected in the middle of the twentieth century. This is also when space exploration and the digital revolution began. Since then, each passing decade has confronted us with an increasingly narrower path on the different choices we could make for a sustainable future, the many directions we could take, and the possible outcomes yet to come into focus. And the scariest feeling of all, possibly the most empowering, too, is that these choices are, for the first time, entirely and consciously in the hands of *one* species. Our choices have come to determine the fate of all of life on Earth and that of its environment.

We took on formidable forces with little knowledge of the consequences. We are sternly reminded today that we live *inside* a closed (eco) system where life and environment evolve in concert. In it, taking care of ourselves requires us to look after the entire system, every action generating a reaction fed back to the whole through biogeological cycling. As the butterfly effect reminds us, if a butterfly flaps its wings in the Amazon forest, it can change the weather half the world away. Abuses to this balance might have been forgiven in the past and absorbed in the background noise of the planet's workings when our population was under a billion individuals. It is impossible today with 8 billion.

The Earth is overcrowded and choked up by human activity. The uncontrolled growth of our civilization is officially consuming more land and sea resources than our planet can produce each year, as has been the case for a few years now. Sustaining our civilization comes at the cost of losing 150 species *per* day,[2] in what is now considered the sixth mass extinction. It also burdens the environment in irreparable ways. We probably can outsmart nature for just a little while longer by engineering food and producing more drinking water through ocean desalination as we lose glaciers globally. A wiser approach would be to make executive decisions to release our grip on our planet and stop assaulting the environment now. The Earth is vast, but it is not expandable. It has finite resources, and we are running out of time to decide which future we care to embrace. Here, the answer is as much about who we are as about who we want to become.

We may wish to continue ignoring our downward spiral of self-destruction, in which case, the outcome is already determined. As we completely relinquish our survival instinct, we will sacrifice not only an entire biosphere but also, most extraordinarily for any species, our own kind.

Changes have made and broken countless species since the dawn of time as part of evolution. That much is true, and probably best exemplified by one of the worst ecological disasters in history that followed the release of oxygen into the Earth's atmosphere by cyanobacteria a few billion years ago. The Great Oxygenation Event killed 90 percent of all life on Earth and altered the environment forever, including our planet's signature in space. But, at least, it benefited cyanobacteria and the rise of aerobic species, creating in the process a uniquely diverse biosphere that, with time, spread all over the world. What benefit could the systematic destruction of our planet and its biodiversity possibly have for our species? And why can't we snap out of this obvious illusion of invincibility and denial in front of ever-intensifying heat and natural disasters? Our footprint and impact on our world become more devastating every day. Yet, we look the other way in a collective delusional posture, leaving an unfamiliar and hostile environment to our children.

We are increasingly disengaged from our natural roots. Our emotional relationship to our planet is no different anymore from our feelings for a shopping mall: We visit episodically for distraction and take what we need from it. It provides entertainment and resources without us being concerned about how the shelves are being restocked or the mess we leave behind for others to clean up. Once satisfied, we return quickly to our artificial bubbles and bury our heads into the sand. Our disengagement is in part fear-driven by the false notion that we, as individuals, are powerless to do anything to stop an unraveling planetary-scale tragedy. We feel small and scared by the magnitude of a monumental challenge and its urgency. Meanwhile, we look at each other, waiting for Superman to save the day and chase the villain. But nobody will come. We are on our own, and each of us is both the superhero and the villain. Through our everyday actions, even the most minute ones, we can increase the power of one over the other. Nothing happens outside our planetary bubble and everything matters. That's the bad news, but it is also an opportunity in disguise because it gives us much more power to fight than we give ourselves credit for, if only we cared to care.

COURSE CORRECTION

The same chain of reactions that led to our current downfall can be reversed by simple daily changes. It might not be spectacular at first and will take time, but as the system builds up, it will rebalance itself. If cyanobacteria could have talked 2 billion years ago, probably none of them would have believed they could change the environmental course of an entire planet. Imagine this: A sixty-micron-large microorganism against mighty planet Earth. Taken individually, this was the ultimate David versus Goliath contest. But, one by one and together, they transformed the Earth completely and redefined a living world. Likewise, we can change our fate at any time if we so desire. We still have a small window of time to reverse the current course. But for this to happen, all of us, individually and collectively, must become activists of our own future and fully engage ourselves in a battle in which the outcome is no less than the survival of our civilization.

But fear is not the only factor at play in this unraveling disaster. Our disengagement may also come from our growing reliance on technology-driven, artificial living conditions and on cyberspace to connect us with, and keep us informed about, the outside world. We often joke of not being able to survive without these tools, but could this be more than a simple jest? If we look at the process from an ecological standpoint, it actually makes perfect sense. For most, the primary stimulus today does not come from the direct interaction with nature anymore. It comes through artificial environments. In many ways, it could be interpreted as the most recent sign of an upcoming adaptation of a species to a new environment. It just happens that this new environment is technological and digital. Elements deemed most critical to survival and competitiveness, information management and responses to changes are now located on these media. In that respect, the vision of a cyborg humanity is a self-fulfilling prophecy of a creator becoming its creation.

It would not be too different from the old cyanobacteria producing a new atmosphere a few billion years ago, except for a vital difference: this time, coevolution has shifted from involving life and its natural environment, which formerly engaged the entire biosphere through bio-geological cycles, to a coevolution of humans and an artificial habitat that only engages humanity. Unfortunately, this balancing act can only be maintained through the ultimate conundrum: the sacrifice of an entire planetary ecosystem to harness the resources necessary for the survival of the one alpha predator species—us. Pushed to the extreme, this process could be likened to a mutation, and in many ways to cancer. Once a cell's messaging has become compromised and part of the illness, its only hope for survival is to see the tumor grow. Being part of the tumor is the new environmental norm with disregard to the fact that the growth endangers the overall body that supports it. However, contrary to cancer, humanity is in a position to reverse the process at any time, or at least, manage it in ways that could maintain natural balance, while still allowing technological progress. But right now, this is not happening, and alarm bells are being met with complacency fueled by short-term interests.

The management of a balance with the natural environment as a primary element of survival is not perceived by our species as critical anymore because it is located outside the technological and digital bubbles that became central to our lives. Before any other, this strikes me as the first singularity, and it is already well underway. In the past few decades, it has transformed our relationship with our environment and our perception and acceptance of what reality is. This is particularly true with our dependence on the digital world, which might have a lot to do with our society's incapacity to react as a unified organism to any crisis anymore, whether it is a pandemic or an environmental emergency. Nowadays, our relationship to each other relies as much, if not more, on digital and virtual connections as on real physical bonds. It is a sort of evolution/involution, in which increased complexity is gained by fragmenting the whole organism that society represents into smaller information-gathering individual cells. Although almost akin to a mitosis, the process is not fulfilled in the biological sense, as information is not self-replicating but instead modified, filtered, or, conversely, not vetted before it is communicated.

If our society were a biological organism, an outside observer would see it as experiencing a state of permanent mutation. In it, signals are constantly being crossed, and a confused brain struggles to understand, enable, and maintain essential functions that, in the case of humanity, are still connected to our planet's vitality, although we act as if they were not. As a result, we are increasingly losing our objective experience of the physical world that has guided our evolution up to recently, replacing it by an indirect and subjective experience where data about the world come through digital and technological filters we do not have any control over most of the time. Then it becomes increasingly harder to decide what reality really means, and what is best for our survival. The fundamental issue with this scenario is that interactions with the physical environment (stimuli) and data collection are absolutely essential factors in decision-making, life's survival, and evolution.

It would be absolutely foolish to blame the digital revolution, technological enhancements, or internet and cyberspace for the crisis we face. They are not the problem. They all bring critical benefits and improvements in our lives. They helped us connect during the pandemic and prevented a complete humanitarian and economic collapse during a critical time. It is not the tools we create that are necessarily bad; it is how we use them and how their educational and humanitarian values come second to their gadgetization and moneymaking potential. What is alarming is their misuse, an abuse that alters our view of reality and reinforces escapist behaviors.

Long before the pandemic locked us down, large segments of the population already spent a substantial amount of time exploring virtual worlds to play and meet in cyberspace through avatars, rather than meeting and socializing in person. In addition to satisfying our natural human curiosity and propensity for exploration, the success of this technology stems from the sense of control it provides over whatever universe, planet, or situation we create and explore in the metaverse. In these virtual environments, we can reset the button and restart the session when anything goes wrong and refresh the scenery as if nothing ever happened. Ultimately, we escape reality in a dimension where we can exercise recklessness without penalty and find oblivion and redemption every time we turn the switch on and off, something we feel powerless to achieve in the real world. In the same way, we dive into our inner universe with enhanced technologies to enjoy broader and altered states of consciousness. As the outside world is closing in on us, we take refuge in the universe within. Unfortunately, our inner bubble is still located within the environmental sphere of our planet. There is really no escape.

Without any course correction to the current trajectory, our fate is sealed. We will leave behind ash and dust where not so long ago large rivers ran, the mighty glaciers of our childhood cascaded down valleys, and the pulsing lake levels measured the passing of seasons. A massive geoengineering of our planet will be necessary to create artificial sinks for greenhouse gases and to feed future generations of humans, when Earth was enough to support the expansion of an entire biosphere for 4 billion years. Our planet of bountiful biodiversity that once occupied all possible habitats is now a world in which many large non-domesticated mammals face extinction, where oceans are depleted, and their life-sustaining currents at risk of shutting down. The coming generations of humans will find themselves shut in for extended periods of time during the year to avoid extreme weather, killer heat, and unbreathable air.

The rise in concentration of carbon dioxide will soon pale in comparison to the added effects of methane gas. Trapped at the bottom of the oceans and in frozen soils for millions of years, it is now released in the atmosphere by the collapse of the permafrost over areas extending millions of square kilometers, and it has more than eighty times the greenhouse effect of CO_2 over the first twenty years after reaching the atmosphere. If left unchecked, increased methane concentrations will trigger a runaway effect leading to a continuous rise in temperatures, which will melt more ice and release more methane. This is one possible future, a humanity that is confined to house arrest on its own planet, when it was ready to spread its wings into space.

INTERSTELLAR

Going to the moon or Mars won't save us, at least not in the short term. Our space technology is nascent, and so is our ability to settle on other worlds. We make extraordinary strides each day, inching closer to a major discovery in propulsion systems that will revolutionize space travel and make all of our goals of planetary expansion possible sooner. Meanwhile, humans are heading back to the moon after an almost seventy-year hiatus. The first settlement will humbly start there, close to home. We might even land a crew on Mars within a decade, but that will not be to stay at first. Critically, all these initial efforts will be highly dependent on our planet for logistics and survival, and for a very long time. Earth was foundational to our past; it is vital to sustain our present and it is critical to seed our future.

Regardless of time and logistics, how tragic would it be for humanity to focus its space ambitions around the idea of escaping a planet it ravaged? Instead, why can't we take in stride Arthur C. Clarke's message in 2010: *Odyssey Two*: "All these worlds are yours." This message resonates with who we are at this juncture because, despite all of our flaws and weaknesses, we have come so far in such a short period of time. In fact, a large part of the problem lies there, too. Think that, only two centuries ago, the fastest way to reach our destination was still the best horse money could buy! We are overwhelmed by exponential knowledge and capabilities that we don't really know how to handle well yet, and we are still at a stage where we feel the intoxication of endless possibilities without thinking too much about the consequences.

We are neither bad nor evil. We are simply a young, restless, inexperienced civilization still learning to use matches to light a fire and coming close to burning the house down in the process. Let's feel the heat and take responsibility. Let's transform the challenge into an opportunity. Let's face the fear and take action. We do not have much time left to do so, but we can act now. There is a vibrant future for humanity as a long-term civilization that understands the balance of nature on its planetary home and can spread this hard-won wisdom across the solar system and beyond.

Despite Arthur C. Clarke's warning, we are even bound to land on Europa in a few years from now. And while we might find life on Jupiter's small moon one day, the monolith won't be waiting for us there, and we won't have to go that far to find it, either. A symbol of profound transitions in human evolution in Clarke's novel, the monolith is here on Earth, right in front of us, casting a deep shadow. We are staring at it everywhere we look, each passing day. Its towering presence imposed itself at the end of the 1950s, announcing a Gaia Super Event, a time of passage when, for the very first time, a species of Earth was given the possibility to master its destiny. Since then, we have often chosen to use this freedom to override nature's cycles and create chaos.

If we fail, nobody will ever know that somewhere at the periphery of the Milky Way, a world once came to life. For what looked like a blink of a cosmic eye, on that world of somewhat insignificant stature, myriads of strange creatures would go on to rhythm their lives to the seasonal heartbeat of their planet. Over eons, changing generations of life-forms of ever-increasing complexity observed their star rise and set over the horizon, bending shapes and skills to adapt to ice and heat as nature relentlessly followed its course. The many colors of our people, the sound of our music, the ingenuity of our science, the beauty of our architecture, and the strength of our love will be all but lost and forgotten as the waves of the cosmic ocean pass, oblivious, over a now silent planetary shore where once birds sang, Mozart played, and Einstein dreamed.

In this probabilistic universe where we still search for evidence of another life, how extraordinary is it that we can even think about all those things? How magnificent of a species are we truly for having gone through the battlefields of 4 billion years of evolution on our Pale Blue Dot in so many different incarnations and still surviving? Countless generations have come and gone, and with them an encrypted code of life remained the unbroken link to each other and to our very first moment on this world. Indeed, we are survivors and we are a precious gift, and such is *every* form of life on Earth, to which we are related through our common origin. Think of how even the most humble terrestrial microorganism we ignore out of contempt would be a treasure of incommensurable meaning if it were found on any other planet. Equally precious is the Earth, which allowed the process of life to take place. This is our heritage; this is our treasure to protect, and our cosmic message to carry and spread afar when, one day, we settle on the moon and Mars, create outposts high up in the atmosphere of Venus, on asteroids, and maybe on Titan and far beyond. But we need to do all of these things not out of desperation, but because we choose to do so as part of humanity's journey, as stepping stones on the way to our interstellar destiny.

Nowhere is it written yet that we will fail, but time is running short. And we should not oppose the care for our planet with our aspirations to reach for the stars because they feed each other in a symbiotic relationship that may hold the key to overcoming the crisis we currently face. Over the last couple of decades, space developments have greatly improved our understanding of Earth's climate. New generations of satellites inform us at the global to local scale, and in near real time, about the state of the atmosphere, the land, and the oceans, and about the evolution of environmental factors. The volume of data they generate facilitates informed decision-making and raises awareness of changes and evolution. Meanwhile, methods developed by astrobiology integrating datasets from global scale to microscale, and from orbit to the ground, designed to search for habitats and life on other planets, may soon provide a new way to reduce vexing forecasting anomalies in climate models here on Earth, and help improve the timing and effectiveness of our mitigation plans. These are powerful weapons we can use for a better future.

Escapist arguments are unnecessary to demonstrate that space exploration is foundational to the future of humanity, starting on our own planet. Beyond the technological benefits, the science behind the study of planetary atmospheres provides a guiding hand to understand systems dynamics, feedback mechanisms, and runaway effects, and how to stay away from irreversible thresholds. In that, the perspective gained through the exploration of Mars, Venus, and Titan have a lot to contribute in helping us monitor the evolution of our own atmosphere. Further, searching for life on other planetary worlds uniquely opens our minds to the critical importance of maintaining the balance between life and environment in order to understand how habitability was lost, or never was, in so many cases. It further shows us how delicate of an exercise this is and provides us with critical data to identify tipping points that cannot be crossed here on Earth.

In the same way fundamental keys to our future on Earth reside in the knowledge we gather exploring alien worlds; essential skills on how to expand humanity beyond our planet and transform hostile planetary environments to make them livable for future humans reside in the knowledge we gather today on Earth while trying to tame an increasingly hostile planet. The solution to our current quandary lies in the same ingenuity that often causes our downfall. We simply need to change our perspective on how to use our inventiveness, and we must teach our children to do the same. We have the science and the technology. We also have the ingenuity, the ideas, and the tools. Let's put them to good use, because there is no room and no time anymore for our obsolete selfish, careless, and predatorial ways. We are also running out of sand in which to bury our heads. Time and the cosmic hourglass will be our mirrors and the ultimate judge of our character.

If we navigate the storm successfully, countless wonders await us on our planet and in the universe. Meanwhile, I am often reminded of a piece of wisdom that comes from my time advocating for the Gusev crater as a landing site for the Spirit rover on Mars. One of the first things we were told in the opening remarks of the first workshop was that our candidate landing sites could feature the best science in the solar system. Yet, if they were unsafe to land on, we were not going. You simply do not have a mission if you crash your spacecraft. Humanity's mission right now should be unwaveringly focused on preserving the integrity of its vessel in space and ensuring a future to all of those under its responsibility. Our children, of course, but as the apex species, we have a responsibility for the Earth and its biosphere as well. Without them, the human odyssey stops here.

This is a moment when we need to answer Sagan's question in *Cosmos*: "Our children long for realistic maps of the future they can be proud of. Where are the cartographers of human purpose?" Throughout history, in their journeys of exploration, humans have sailed into the unknown on Earth's oceans, often navigating by the stars, with no maps, but just their gut instinct telling them that something was over the horizon. This goal gave them purpose, and it gave meaning and a sense of direction to their voyage. Cartographers established the maps as they went, while opening new waterways. They made the journey easier for the next travelers and, in time, transformed uncharted territories into trade routes. Today, we struggle in rough seas, trying to navigate a storm of hellish proportions on a changing planet. If our purpose is to endure the passage of time as a civilization, our goal then becomes crystal clear: Let's take care of the Earth, our vessel in space, and use all the tools and ingenuity we have as the many guiding stars on our journey into the darkness of the unknown, wherever their light may be shining from, whether it is through our understanding of nature's cycles, education, technology, science, or space exploration. May this illumination guide us to safe cosmic shores, while we map the course of new routes for the next generation to navigate. May today's uncertain voyage help us redefine what humanity truly means and this time be remembered as the age when humans rose to the challenge of becoming planetary citizens.

ACKNOWLEDGMENTS

Frank D. Drake
(May 28, 1930–September 2, 2022)

On September 2, 2022, the astronomer Frank Drake passed away. That day, the search for life in the universe lost one of its pioneers, and SETI a founding father. From a personal standpoint, barely two weeks after Edmond's passing, Frank's death left me orphaned of another discrete but central presence in my life, this time on a professional level. It was Frank who reviewed my application to the SETI Institute in 1998 and accepted it a few days later. I was privileged to meet and talk to him often at the institute, and even more often just delighted to listen. Frank was humble, funny, and had a boundless intelligence. Despite the passing of time, he had kept a spark in his eyes that betrayed his ever present passion and vision. In the years following my nomination as the chief scientist at the SETI Institute's Carl Sagan Center, he kept supporting and encouraging me. He leaves behind a tremendous scientific heritage, a springboard to the stars for the generations to come, a vision that opens to endless cosmic horizons, a vision each day more relevant.

Thinking about both Edmond and Frank and their creative energy gave me the resolve to finish this manuscript after their passing. And, in a merciless year, I also found treasures of friendship and comfort in my colleagues, friends, family, and the public—and I am forever grateful to them.

I also want to especially thank all of those who went out of their way and took time to read the first draft of the manuscript and provided positive comments that helped improve it. So, thank you, Rose Orenstein, my first reader and guardian of the English preposition! Thank you, Rose, for your wonderful heart and unwavering friendship.

My SETI Institute colleagues also showed that team spirit that epitomizes them so perfectly and they dived into the first draft as well. Special thanks to Rebecca McDonald (communications director), Simon Steel (deputy director, Carl Sagan Center), and Bill Diamond, our president and CEO. The conversations Bill and I regularly have on these subjects that fascinate us both, of life beyond Earth and the future of our planet, have certainly fed my reflections and the thoughts shared in these pages. (Bill is also a talented photographer and I want to thank him for the picture of the ATA he provided that is included in these pages.)

Beyond this book, I would like to thank my team, all my colleagues at the SETI Institute, NASA Ames, and the broader NASA family for their unconditional support during one of the most difficult times of my life.

It goes also without saying that this book would not exist if not for the outstanding professionalism of the LE SEUIL team. I discovered the scope of their talent, commitment, and kindness with *Voyage*, my first book. Thank you, Muriel, for being the little fairy that always smooths corners so discreetly and efficiently, and always with a smile; Severine, the genius agenda and logistics juggler for interviews and media, with whom I rediscovered Paris by taxi (in all directions) in 2021; Josephine, for her laser vision in reading over the manuscript and her pertinent questions that improved it. My gratitude also goes to Laurent, for our discussions, your visions of Hollywood for my works, and for being my guide in a world that sometimes looks like a parallel universe to the astrobiologist I am.

I would like to extend a very special thank-you to Adrien for helping me shape my manuscript, like a sculptor with clay. I am deeply grateful for your attention to details, your pursuit of excellence . . . and your patience with my anglicisms in the French version!

I am also deeply grateful for the support of the team at Scribner who have worked tirelessly to make the publication of this book possible in the US and UK, those I have interacted with on a regular basis and also those who made things happen behind the scenes. Thank you Chris Richards, executive editor, and Kris Doyle, deputy publishing director, for your enthusiasm, kind words and your guidance always; Joie Asuquo, for helping me with the editorial process and making sure we stayed on schedule; I am in debt to Nan Graham and Stu Smith, publisher and associate publisher at Scribner, respectively, for believing in this book and letting people know! Thank you, Dan Cuddy, for overseeing the copy-editing, proofreading, and the fine-tuning at the various stages of the manuscript. And thank you, Kate Kenney-Peterson, for designing the interior of the book. To all of you, my deepest gratitude for making me feel so welcome in the Scribner family.

Finally, thank you, Merlin. Words cannot even come close to describing how much I miss you. After thirty-six years together, this manuscript is the first one you did not finish reading and commenting on. I hope you will still be proud of me. I finished it thinking of you.

IMAGE CREDITS

103   NASA/JPL-Caltech/SSI/Cassini Imaging Team/Jason Major.

107   NASA/JPL-Caltech. Artist illustration.

117   Left image: NASA/JPL/Space Science Institute; right image: ESA/NASA/University of Arizona.

119   NASA/JPL-Caltech/ASI/USGS: PIA17655.

127   NASA/JPL-Caltech/University of Arizona/LPG Nantes. Cassini mission (mapping in the visible and infrared, VIMS). Picture No. PIA20022.

129   Johns Hopkins University Applied Physics Laboratory.

133   NASA/JPL-Caltech/UCLA/MPS/DLR/IDA/PSI. PDS Image ID: PIA22480.

137   NASA/John Hopkins University Applied Physics Laboratory/ Southwestern Research Institute/National Optical Astronomy Observatory. LORRI/New Horizons.

140   NASA/Johns Hopkins University Applied Physics Laboratory.

144   NASA/JHUAPL/SwRI.

149   NASA/Johns Hopkins University Applied Physics Laboratory/ Carnegie.

156   Graphic and legend: NASA Ames.

159   NASA/ESA/D. Kipping (Columbia University), et A. Feild (STScI).

163   ESO, Paranal Observatory.

177   ESO/K. Meech et al.

185   NASA/JPL-Caltech.

213   Kevin M. Gill.

217   Bill Diamond.

218   Eliot Gillum.

222   NASA/JPL-Caltech.

232   Image: NASA.

265   Seth Shostak, SETI Institute.

PHOTO INSERT

Overview effect. Credit: NASA-ISS035-E-34689.

Terrestrial biosignatures. Source: NASA—Landsat 8, NASA Earth Observatory images by Joshua Stevens and Lauren Dauphin, using Landsat data from the U.S. Geological Survey and MODIS data from LANCE/EOSDIS Rapid Response.

Comet 67P/Churyumov-Gerasimenko. Source: ESA/Rosetta/MPS for OSIRIS Team MPS/UPD/LAM/IAA/SSO/INTA/UPM/DASP/ IDA; processing by Giuseppe Conzo.

The surface of Venus seen from the Venera 9 probe on October 22, 1975. Source: Lavochkin, Union of Soviet Socialist Republics. Image processing: Ted Stryk.

Sunrise over the Meridiani Planum on Mars. Source: NASA/JPL-Caltech/Cornell.

Martian delta landscape. Source: NASA/JPL-Caltech/ASU/MSSS.

The slopes of Mount Sharp. Source: NASA/JPL-Caltech/MSSS/fredk/ S. Atkinson.

An ocean of dunes. Source: NASA/JPL-Caltech/ASU.

Ganymede. Source: NASA/JPL-Caltech/SwRI/MSSS. Image processing: Thomas Thomopoulous.

Europa. Source: NASA/JPL-Caltech/SwRI/MSSS. Image processing: Navaneeth Krishnan S © CC BY.

Enceladus. Source: NASA/JPL-Caltech/SSI.

Polar Vortex on Titan. Source: NASA/JPL-Caltech/SSI.

Titan's "Magic Island." Source: NASA/JPL-Caltech/ASI/Cornell (PIA20021).

The Occator crater. Credit: NASA/JPL-Caltech/UCLA/MPS/DLR/ IDA: Nico Schmedemann, Guni Thangjam, and Andreas Nathues of the Dawn Framing Camera Team.

Pluto's atmosphere. Source: NASA/JHU-APL/SwRI. Image processing: Roman Tkachenko.

The frozen canyons of Pluto's north pole. Source: NASA/Johns Hopkins University Applied Physics Laboratory/Southwest Research Institute.

James Webb Space Telescope's first deep-field image. Source: NASA, ESA, CSA, STScI, NASA-JPL, Caltech.

At the heart of the Ghost Galaxy. Source: NASA/ESA/CSA, J. Lee and The PHANGS-JWST team.

Cosmic cliffs. Source: NASA/ESA/CSA/STSci.

Planetary nebula NGC2022. Source: ESA/Hubble, NASA, R.Wade.

Protoplanetary disks captured by ALMA. Source: ALMA, ESO, NAOJ, NRAO, S. Andrews, N. Lira.

HIP 65426b. Credits image: NASA, ESA, CSA, Alyssa Pagan (STScI). Credits science: Aarynn Carter (UC Santa Cruz), ERS 1386 Team

HR8799. Source: NASA/JPL-Caltech/Mount Palomar Observatory.

Apep star system. Source: ESO, Joseph R. Callingham.

The birth of a planet (1514 PDS). Source: ESO/A. Muller et al.

Betelgeuse. Source: ESO/ALMA/P. Kervella.

Comet 2l/Borisov. Image source: NASA/ESA/David Jewitt. Image processing: Joseph DePasquale (STSci).

Hubble and JWST observe the impact of DART on Dymorphos. Sources: Science—NASA/ESA, Jian-Yang Li (PSI), Cristina Thomas (Northern Arizona University), Ian Wong (NASA-GSFC). Image processing: Joseph DePasquale (STSci), Alyssa Pagan (STSci).

The Black Marble. Source: NASA Earth Observatory.

NGC 346. Source: NASA, ESA, CSA, O. Jones (UK ATC), G. De Marchi (ESTEC), and M. Meixner (USRA), with image processing by A. Pagan (STScI), N. Habel (USRA), L. Lenkic (USRA) and L. Chu (NASA/Ames).

The Pillars of Creation (JWST NIRCam). Source: Science—NASA, ESA, CSA, STScI. Image processing: Joseph DePasquale (STScI), Anton M. Koekemoer (STScI), Alyssa Pagan (STScI). Webb's NIRCam was built by a team from the University of Arizona and Lockheed Martin's Advanced Technology Center.

NOTES

CHAPTER 1: THE OVERVIEW EFFECT

1. D. S. Spiegel et al., "Generalized Milankovitch Cycles and Long-Term Climatic Habitability," *Astrophysical Journal* 721, no. 2 (September 2010): 1308–18.

CHAPTER 2: SPARKS OF LIFE

1. Stephen R. Meyers and Alberto Malinverno, "Proterozoic Milankovitch Cycles and the History of the Solar System," *Proceedings of the National Academy of Sciences* 115, no. 25 (June 2018): 6336–68.
2. Ryo Mizuuchi, Taro Furubayashi, and Norikazu Ichihashi, "Evolutionary Transition from a Single RNA Replicator to a Multiple Replicator Network," *Nature Communications* 13, no. 1 (March 2022): 1460.
3. Sean F. Jordan et al., "Promotion of Protocell Self-Assembly from Mixed Amphiphiles at the Origin of Life," *Nature Ecology and Evolution* 3 (November 2019): 1705–14.
4. Bruce Damer and David Deamer, "The Hot Spring Hypothesis for an Origin of Life," *Astrobiology* 20, no. 4 (December 2019): 429–52.
5. Jeremy England, *Every Life Is on Fire: How Thermodynamics Explains the Origins of Living Things*, illustrated ed. (New York: Basic Books, 2020), 272 pages (11–12).

6. Christopher P. Kempes and David C. Krakauer, "The Multiple Paths to Multiple Life," *Journal of Molecular Evolution* 89 (July 2021): 415–26.

7. C. E. Cleland, "Epistemological Issues in the Study of Microbial Life: Alternative Biospheres," *Studies in the History and Philosophy of Biological and Biomedical Sciences* 38 (2007): 847–61.

8. P. C. W. Davies, S. A. Benner, C. E. Cleland et al., "Signatures of a Shadow Biosphere," *Astrobiology* 9(2) (2009): 241–49.

CHAPTER 3: VENUS AND ITS VEIL OF SECRECY

1. Jane S. Greaves et al., "Phosphine Gas in the Cloud Decks of Venus," *Nature Astronomy* 5 (2021): 655–64.

2. N. Truong and J. I. Lunine, "Volcanically Extruded Phosphides as an Abiotic Source of Venusian Phosphine," *Proceedings of the National Academy of Sciences* 118, no. 29 (July 2021): e2021689118.

3. Haygen Warren, "Exploring VERITAS, One of NASA's New Missions to Venus," NASASpaceFlight, July 24, 2021, https://www.nasaspaceflight.com/2021/07/veritas-nasa-venus/.

4. William Steigerwald and Nancy Neal Jones, "NASA to Explore Divergent Fate of Earth's Mysterious Twin with Goddard's DAVINCI+," NASA, June 2, 2021, https://www.nasa.gov/feature/goddard/2021/nasa-to-explore-divergent-fate-of-earth-s-mysterious-twin-with-goddard-s-davinci.

5. ESA/NASA, EnVision: Understanding Why Earth's Closest Neighbor Is So Different website, https://envisionvenus.eu/envision/.

6. Janusz Petkowski and Sara Seager, "Venus Life Finder Mission Study," Massachusetts Institute of Technology, December 10, 2021, https://venuscloudlife.com/venus-life-finder-mission-study/.

CHAPTER 4: BLUE SUNSETS

1. H. P. Klein, "The Viking Biological Experiment," *Icarus* 34, no. 3 (1978): 666–74.
2. "Airbus to Bring First Mars Samples to Earth: ESA Contract Award," Airbus, October 14, 2020, https://www.airbus.com/en/newsroom /press-releases/2020-10-airbus-to-bring-first-mars-samples-to-earth -esa-contract-award.
3. Description of possible life on Mars adapted from Nathalie A. Cabrol, "Tracing a Modern Biosphere on Mars," *Nature Astronomy* 5, no. 3 (March 2021): 210–12.

CHAPTER 5: PLANETARY SHORES

1. Gaël Choblet et al., "Powering Prolonged Hydrothermal Activity Inside Enceladus," *Nature Astronomy* 1 (2017): 841–47.
2. Masahiro Ono, "Enceladus Vent Explorer," NASA, April 7, 2020, https:// www.nasa.gov/directorates/spacetech/niac/2020_Phase_I_Phase _II/Enceladus_Vent_Explorer/.
3. Ethan Schaler, "SWIM—Sensing with Independent Micro-Swimmers," NASA, February 19, 2022, https://www.nasa.gov/directorates/space tech/niac/2022/SWIM/.
4. Samuel M. Howell and R.T. Pappalardo, "NASA's Europa Clipper—a Mission to a Potentially Habitable Ocean World," *Nature Communications* 11, no. 1 (March 2020): 1311.

CHAPTER 6: TITAN: A WORLD OF UNKNOWNS

1. Robert H. Brown, Jean-Pierre Lebreton, J. Hunter Waite, eds., *Titan from Cassini-Huygens* (Dordrecht, Netherlands: Springer, 2010), 535 pages.

2. Rosaly M. Lopez et al., "Titan as Revealed by the Cassini Radar," *Space Science Reviews* 215, no. 4 (May 2019): https://doi.org/10.1007/s11214-019-0598-6.

3. Shannon MacKenzie et al., "Titan: Earth-Like on the Outside, Ocean World on the Inside," *Planetary Science Journal* 2 (March 2021): 112.

4. Ralph D. Lorenz et al., "Dragonfly: A Rotorcraft Lander Concept for Scientific Exploration at Titan," *Johns Hopkins APL Technical Digest* 34, no. 3 (2018): https://dragonfly.jhuapl.edu/News-and-Resources/docs/34_03-Lorenz.pdf.

CHAPTER 8: REVOLUTIONS IN THE NIGHT SKY

1. Ryan Oliver, *Aliens and Atheists: The Plurality of Worlds and Natural Theology in Seventeenth Century England*, Master of Arts thesis, University of North Texas, December 2007.

2. Michael Crowe, ed., *Extraterrestrial Life Debate: Antiquity to 1915* (Notre Dame, IN: University of Notre Dame Press, 2008).

3. Nathalie A. Cabrol, "Beyond the Galileo Experiment," *Nature Astronomy* 3 (July 2019): 585–87.

CHAPTER 10: ECHOES OF COSMIC WAVES

1. Charles River Editors, *The Fermi Paradox: The History and Legacy of the Famous Debate over the Existence of Aliens* (Ann Arbor, MI: Charles River Editors, 2020).

2. D. Linotte, "The Duration of Civilizations—Preliminary Recalculations (A Short Note), *International Journal of Humanities and Social Science Invention (IJHSSI)* 9, no. 7 (2020): 10–13.

3. Peter D. Ward and Donald Brownlee, *Rare Earth: Why Complex Life Is Uncommon in the Universe* (Göttingen, Germany: Copernicus, 2003).

4. Paul Davies, *The Eerie Silence: Renewing Our Search for Alien Intelligence*, 1st ed. (New York: Houghton Mifflin Harcourt, 2010).

5. Caleb Schaft, "Is Physical Law an Alien Intelligence?" Nautilus, November 11, 2016, https://nautil.us/is-physical-law-an-alien-intelligence -236218/.

CHAPTER 11: CONNECTING BLUE DOTS

1. "Declaration of Principles Concerning the Conduct of the Search for Extraterrestrial Intelligence," unanimously adopted by the SETI Permanent Study Group of the International Academy of Astronautics, at its annual meeting in Prague, Czech Republic, on September 30, 2010, SETI Institute, https://www.seti.org/protocols -eti-signal-detection-0.

2. Seth Shostak, "The UAP Story: The SETI Institute Weighs In," SETI Institute, June 25, 2021.

3. "Preliminary Assessment: Unidentified Aerial Phenomena," unclassified report, Office of the Director of National Intelligence, June 25, 2021, https://www.dni.gov/files/ODNI/documents/assessments /Preliminary-Assessment-UAP-20210625.pdf

4. "Unidentified Aerial Phenomena," NASA, 2023, https://www.nasa.gov /feature/faq-unidentified-aerial-phenomena-uapsufos.

5. "The UAP Story: The SETI Institute Weighs In," SETI Institute, June 25, 2021, https://www.seti.org/press-release/uap-story-seti -institute-weighs.

CHAPTER 12: PARADOXES, PARADIGMS, AND THE GRAMMAR OF LIFE

1. Marc Neveu et al., "The Ladder of Life Detection," *Astrobiology* 18, no. 11 (November 2018): 1375–1402.

2. Eric Smith and Harold J. Morowitz, *The Origin and Nature of Life: The Emergence of the Fourth Geosphere*, 1st ed. (Cambridge, UK: Cambridge University Press, 2016).

3. Stephane Tirard, Michel Morange, and Antonio Lazcano, "The Definition of Life: A Brief History of an Elusive Scientific Endeavor," *Astrobiology* 10, no. 10 (December 2010): 1003–9.

4. Matthew W. Powner, Béatrice Gerland, and John D. Sutherland, "Synthesis of Activated Pyrimidine Ribonucleotides in Prebiotically Plausible Conditions," *Nature Letters* 459 (2009): 239–42.

5. Lisa Hays, ed., *NASA Astrobiology Strategy 2015*, NASA, 2015, https://astrobiology.nasa.gov/nai/media/medialibrary/2015/10/NASA_Astrobiology_Strategy_2015_151008.pdf.

6. Jacques Monod, *Chance and Necessity: An Essay on the Natural Philosophy of Modern Biology* (New York: Vintage, 1972).

7. J. Baross et al., *The Limits of Organic Life in Planetary Systems* (Washington, DC: National Academies Press, 2007).

8. David Grinspoon, *Earth in Human Hands: Shaping Our Planet's Future* (New York: Grand Central, 2016).

9. Gerard Jagers op Akkerhuis, "Explaining the Origin of Life Is Not Enough for a Definition of Life," *Foundations of Science* 16, no. 4 (November 2011): 327–29.

10. Edward N. Trifonov, "Vocabulary of Definitions of Life Suggests a Definition," *Journal of Biomolecular Structure and Dynamics* 29, no. 2 (October 2011): 259–66.

11. Ryo Mizuuchi, Taro Furubayashi, and Norikazu Ichihashi,

"Evolutionary Transition from a Single RNA Replicator to a Multiple Replicator Network," *Nature Communications* 13, no. 1 (March 2022): 1460.

12. Erwin Schrödinger, *What Is Life? The Physical Aspect of the Living Cell.* Based on lectures delivered under the auspices of the Dublin Institute for Advanced Studies at Trinity College, Dublin, in February 1943, https://chsasank.com/classic_papers/what-is-life-physical-aspect -of-living-cell.html.

13. Jeremy England, *Every Life Is on Fire: How Thermodynamics Explains the Origins of Living Things*, illustrated ed. (New York: Basic Books, 2020).

14. Daniel A. Fiscus, D. A. (2002): "The Ecosystemic Life Hypothesis III: The Hypothesis and Its Implications," *Bulletin of the Ecological Society of America* 83, no. 2 (January 2002): 146–49.

15. James E. Lovelock and Lynn Margulis, "Atmospheric Homeostasis by and for the Biosphere: The Gaia Hypothesis." *Tellus.* Series A. Stockholm: *International Meteorological Institute* 26, no. 1–2 (1974): 2–10.

16. Christopher P. Kempes and David C. Krakauer, "The Multiple Paths to Multiple Life," *Journal of Molecular Evolution* 89, no. 7 (2021): 415–26.

17. Borna Jalšenjak, "The Artificial Intelligence Singularity: What It Is and What It Is Not," 107–116, in *Guide to Deep Learning Basics*, ed. Sandro Skansi (Cham, Switzerland: Springer, 2020).

18. Sam Kriegman et al., "Kinematic Self-Replication in Reconfigurable Organisms," *Proceedings of the National Academy of Sciences* 118, no. 49 (December 2021): e2112672118.

19. Louis A. Del Monte, *The Artificial Intelligence Revolution: Will Artificial Intelligence Serve Us or Replace Us?* 1st ed. (Louis A. Del Monte, 2014).

20. Robert Lanza and Bob Berman, *Biocentrism: How Life and Conscious-ness Are the Key to Understanding the True Nature of the Universe* (Dallas, TX: BenBella Books, 2009).

EPILOGUE: TIME AS A COSMIC MIRROR

1. Nerilie J. Abram et al.,"Early Onset of Industrial-Era Warming across the Oceans and Continents," *Nature* 536, no. 7617 (2016): 411–18.
2. Gerardo Ceballos et al., "Accelerated Modern Human-Induced Species Losses: Entering the Sixth Mass Extinction," *Science Advances* 1, no. 5 (June 2015): e1400253.

OTHER RESOURCES

NASA Exoplanet Exploration: https://exoplanets.nasa.gov

European Space Agency's "Cosmic Vision": https://www.esa.int/Science_Exploration/Space_Science/ESA_s_Cosmic_Vision

SETI at Berkeley: https://seti.berkeley.edu/people/Dan.html

SETI Institute: https://www.seti.org

INDEX

INDEX

Kerberos (moon of Pluto), 138
Klein, Harold P., 68
Knutson, Heather, 163
Koch, Dave, 165
Krakauer, David, 43, 242
Kuiper, Gerard P., 112
Kuiper Belt, 136, 142–46, 147, 149

Labeled Release (LR) experiment,
  68, 69, 70
Lagrange points, 177
Lane, Nick, 40
Laplace, Pierre-Simon, 4
Large Synoptic Survey Telescope,
  168, 188
Las Campanas (Chile)
  observatory, 160
Late Heavy Bombardment, 16,
  29–30
Latham, David, 160
Leavitt, Henrietta Swan, 13
Levin, Gilbert, 68
life
  as adaptive information, 43
  artificial intelligence (AI) as
    form of, 242, 243–44
  coevolution with environment,
    17, 24, 26, 201, 241, 242, 263
  conditions for, 7
  consciousness and, 248–50
  definition of, 233, 239–42

Drake equation and search for
  complex life, 190–96
elementary compounds of, 14
"Ladder of Life Detection,"
  234–39
new candidates in search for,
  145–47
prevalence in the universe, 2
search for extraterrestrial life, 5,
  8, 14–17, 26, 42
synthetic, 8
"tree of life," 30, 151
vulnerability of, 11
life, origins of, 2, 5, 6, 7, 16, 27–30,
  238
  biochemical theories, 35–42
  "biological event horizon" and, 90
  conditions for inception of life,
    38–42
  early history of Mars and, 76
  Earth's moon and, 30–33, 148
  human identity and, 251
  Mars as possible source of life
    on Earth, 88–90
  new concepts about, 42–44
  panspermia (life from space)
    theory, 33–35
  paradigm-changing discoveries
    and, 44–46
Lingam, Manasvi, 176
Liu, Cixin, 208

solar system (*cont.*)
  search for life in, 26
  unique position of Earth in,
    46
  See also sun
solar wind, 57, 78
Soviet Union, 71
space, vacuum of, 34, 94
space exploration, 5, 101, 264
  infancy of, 91
  Kepler mission as
    transformational moment in,
    164
  search for extraterrestrial life
    and, 239
  space travel, 203, 224, 259
SpaceX, 62
spectroscopy, 163, 167
Spirit rover, on Mars, 263
Spitzer Space Telescope, 162, 163,
  184
Sputnik satellite, 2, 196
stars
  brown dwarfs, 160, 174, *183*
  K-type, 201, 207
  life cycle of, *41*
  main-sequence, 161, 183
  mass of, 16
  M dwarfs, 201
  Population I, 16
  Population II, 16, 205

Population III, 15–16, 205
red dwarfs, 183, 222
white dwarfs, 180, 183, 184
stars, sun-like, 15, 191, 196
  advanced civilizations around,
    199
  habitable zone of, 14, 165, 194,
    200
  life-span of, 197–98
  number in Milky Way galaxy,
    194
  percentage as hosts of habitable
    worlds, 173
  SETI projects and, 219
  sun-like, 196
Star Wars film series, 179
Steffl, Andrew J., 138
Strelley Pool (Australia) fossils,
  29
Styx (moon of Pluto), 138
Suitcase SETI (spectrum
  analyzer), 220
sulfur, 14, 83, 109, 119
sulfur dioxide, 50, 54
sun
  as average-sized star, 13
  Earth's orbit around, 20
  future as red giant, 145
  habitability of Earth and,
    17
  lifespan of, 18